A
YEAR
of the
STARS

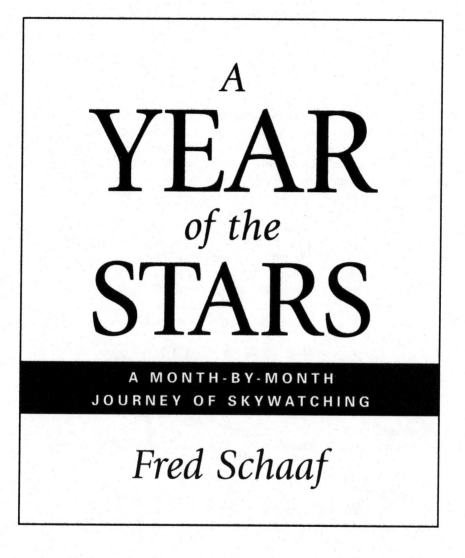

A

YEAR

of the

STARS

**A MONTH-BY-MONTH
JOURNEY OF SKYWATCHING**

Fred Schaaf

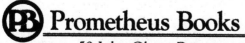

Prometheus Books

59 John Glenn Drive
Amherst, New York 14228-2197

Published 2003 by Prometheus Books

Inquiries should be addressed to
Prometheus Books
59 John Glenn Drive
Amherst, New York 14228–2197
VOICE: 716–691–0133, ext. 207
FAX: 716–564–2711
WWW.PROMETHEUSBOOKS.COM

07 06 05 04 03   5 4 3 2 1

Library of Congress Cataloging-in-Publication Data

Schaaf, Fred.
    A year of the stars : a month-by-month journey of skywatching / by Fred Schaaf.
        p. cm.
    Includes index.
    ISBN 1–59102–092–1 (cloth : alk. paper)
    1. Astronomy—Observers' manuals. I. Title.

QB64.S435 2003
522—dc21

                                                                    2003008108

Printed in the United States on acid-free paper

Contents

CONTENTS

Acknowledgments

This book may never have gotten off the ground were it not for the enthusiasm and support of the incomparable Linda Regan. Linda is an excellent editor, and I want her to know that I greatly appreciated her patience and guidance in the first stages of getting this book written.

A book needs more than a writer and editor; it needs a production manager. Christine Kramer strikes me as being a masterful one, the kind who immediately inspires confidence. I immediately believed that the book would be in able hands with her, and I was right.

I would also like to thank art director Jackie Cooke and other skilled people at Prometheus. Outstanding among these has been Peggy Deemer, the editor who saw this book through its final phases. Those phases are absolutely crucial, and I was fortunate to have such a capable editor as Peggy working with me on them.

Last but far from least in these acknowledgments are the friends and colleagues who contributed to this book. The coverage of double stars in these pages owes much to my fellow astronomy writer James Mullaney, a good man who also is arguably the world's most experienced authority on the deep sky. Doug Myers, great friend and artist for many of my past books, came through again with excellent illustrations. Finally, the biggest contribution to this present book was that of my longtime friend and fellow writer/skylover Guy Ottewell. No matter how busy he has been, Guy has always come to my rescue with his great work and his moral support. In this case, the work is marvelous maps, which, if we've been able to reproduce them as well as they deserve, should be one of the book's outstanding assets.

Introduction

One look at a starry country sky. One listen to amateur astronomers talking about their telescopes and the wonders they've seen.

Either of these experiences may be enough—enough to wake in you the desire to seek the sights of the sky yourself, to learn the high paths of adventure in the heavens that make up the science, hobby, avocation, and passion that is astronomy.

Or either of these experiences may be enough to scare you. How do you learn the identities of all those stars and constellations, and when they appear? What is a planetary nebula or a globular star cluster or M31—where do you find them and how do you tell them apart? There seem to be a multitude of objects and facts to get straight in this hobby. Astronomy is about things which are as far from being mundane (that is, of this world, Earth) or everyday (though they are everynight) as is possible. They are tantalizing, alluring . . . but can you make the leap to understanding and recognizing all these things? You may be simultaneously attracted to and fearful of the bewildering strangeness and richness of astronomy.

If you are afraid, however, fear no more. Astronomy has a reputation for being arcane—that is, known or knowable only to one who has the key. But there is a key, available for everyone. For, as this book explains, there is a simple device which you can use ever so painlessly to teach yourself all astronomy has to offer. The name of this device is the *year*. And all you have to do is let it show you, night by night, and month by month, what the sky has to offer. The learning will mostly come naturally.

Welcome to *A Year of the Stars*—and the year of stars which it chronicles. This book is a guide to letting the heavens teach themselves to you. The secret of learning astronomy is to begin with one night and one part of the sky or even just one constellation—and then let the passing nights and months of the year bring you more and more to add to your initially modest framework of knowledge and sights. Let the year do the work! It sounds like a long period of time (though if you keep finding yourself ready to move on to more, you can always do so by staying up a little later in the night to view the sights which will fill the evening sky in the months ahead). But if you plan to live for at least a few more years, you aren't going anywhere that the heavens won't be. And they aren't going anywhere that you won't be either.

Of course, things are not *quite* as easy as all that. The main section of this book is a month by month tour of hundreds of the best bright stars, constellations, double stars, variable stars, star clusters, nebulas, galaxies, and other celestial sights that can be seen with instruments ranging from the naked eye up to a medium-size (or, in a few cases, large) amateur telescope. But if you have picked up this book as a novice to astronomy then you will need to read the first two chapters which make up an introductory section: a section called "A Primer for New Astronomers." I believe that you will find this section easy and fun (even veteran amateur astronomers may wish to read it to refresh their knowledge and gain perhaps some new perspectives). In fact, I think you will find it downright delectable when you remind yourself that what comes after it is the month chapters which because of this primer you will actually be able to fully understand and follow. And what comes after the current month chapter (an hour or two's read) are the current sights that are waiting out there in that grand universe for you to visit this very night.

You can read the whole book through from start to end, from first month to last. Or you can begin with the month that is currently on the calendar on your wall. Each month chapter is essentially independent of the others. Of course, if you are a beginner you may find yourself sometimes checking back to the book's primer to remind yourself of the meaning of a term or the nature of a kind of celestial object. That is perfectly okay. So, too, is checking the glossary you'll find at this book's end.

I'd like to say a few more words about the maps and tables in this book. The major maps are in a pocket in the back inside cover of the book. For each month there is provided a map which shows the entire starry sky as you would see it in the middle of evenings in that month. It helps you identify where the constellations you want to visit are in the sky. A more detailed map is given for each of the four seasonal sections of the book. These maps show the featured constellations of the season in much greater detail with numerous stars and a number of the most important other celestial objects identified. The positions of deep-sky objects mentioned in the text but not plotted on the maps are given in appendix 5. You can use this information to find them in a star atlas.

Let me stress that this book doesn't provide any detailed discussion on the selection and use of telescopes. That is a topic of its own which you can and should pursue before and after you buy a telescope (recommendations for further reading on this topic are listed in the Sources of Information).

By the way, there are a great number of sights in this book for people who have only their unaided eyes to use: constellations, brightest stars, wide double stars, bright variable stars, the naked-eye Milky Way, annual meteor showers, the Northern Lights, the zodiacal light, and much more. Nevertheless, most people who become seriously interested in astronomy either have a telescope or are eager to get one. Most of the objects described in *A Year of the Stars* are best seen (in some cases, only seen) with a telescope. Indeed, if you are looking for an excellent introductory tour to the heavens for telescopes, I believe that this book will serve your need well. As a matter of fact, I think that even the experienced deep-sky observer will find here at least a few objects he or she hasn't previously encountered and hopefully more than a few new perspectives on—or handy routes for getting to—those objects. In particular, I have tried to present a lot of double stars and facts about them, for this wonderful class of celestial wonder has remained underrepresented for many years in most guides to great deep-sky sights.

In conclusion, I want to note that there is a major feature of this book which I think is unique to a guide of this sort. I am referring to the emphasis on enjoying the entire natural environment in which your astronomical observations take place. I truly love the year and its seasons, as much as I love the sky. The maps and verbal coverage of starry sights in this book is geared toward observers who live at middle northern latitudes—between about 25° and 50° North latitude, though there is much of value here for readers located anywhere in the world. My comments and reflections on both the practical and the aesthetic elements of the observing environment of each month are mostly those of an observer located in the eastern United States—for that is what I am, and have been all my life. Yet many of those elements of the environment are similar to those which occur in most mid-northern latitude lands. And even if the place you live and look from has weather and landscapes and flora and fauna radically unlike those of the eastern U.S., my comments can provide you with an example to follow in noting the nuances of your own observing environment throughout your own year.

Your own year. This is a vital point which *A Year of the Stars* is based on: the year and the heavens and the wonders within them are not unapproachable and they are not someone else's. They don't belong to just someone who owns a fancier telescope or has thousands of dollars of auxiliary equipment or a doctorate in astronomy. They belong to you—to every one of us who has eyes that see and the curiosity to look. It's true that the wasteful component of society's artifical lighting—light pollution—has been stealing much of the night sky which is the natural heritage of us all. But we can all do at least a little to try changing that (see chapter 1 and www.darksky.org). And there are still a lot of starry heavens left for the ever-circling years to give us. Go out and receive them.

PART 1

A Primer for New Astronomers

Chapter 1
Learning and Seeing the Heavens

Whenyou look up into a clear night sky, the wonders you behold are the very ones which have stirred every previous generation of human minds and hearts since before the beginning of history. What do you see? A lustrous slice of Moon? The bright and almost perfectly steady-shining beacon of a proud planet? One thing you will certainly want to see on any clear night is the flickering fires of stars in their awesomely enduring patterns.

It is the starry heavens beyond our solar system which is the focus of this book. But if you are new to astronomy, there are fundamentals you need to know before you can use the chapters about each month's view of the stars. Therefore, this initial two-chapter section of the book is called "A Primer for New Astronomers."

Chapter 1 limits itself to sights which can be seen with the naked eye. But it is not just sights like stars and constellations, meteors and Northern Lights and the Milky Way band which we'll examine. The gist of this chapter is explaining two major sets of topics. One is how astronomers measure positions, brightnesses, angular distances, and time in the sky, and how they have arranged the constellations and given names and designations to the stars. (Without this information, you wouldn't even be able to find your way around the heavens.) The other major set of topics is how to get the most out of your skies and your eyes. This includes not only how to try to avoid great impediments to stellar observing like light pollution, bright moonlight, and haze (not to mention clouds), but also how to use observational techniques which enable you to see more in the heavens.

The starry heavens also include hazy patches of glowing light called *nebulas*; collections of stars called *star clusters*; and much vaster congregations of usually many billions of stars called *galaxies*. A few examples of each of these can be glimpsed with the unaided eye. But the vast majority of such *deep-sky objects*—objects beyond our solar system—can be seen only with the use of a telescope. A telescope or binoculars is also needed if we are to split most examples of close-together pairs of stars called *double stars* or to observe most of the stars which exhibit fluctuations in brightness and are called *variable stars*. Consequently, observational details about these objects and the basics of what scientists have learned about their physical nature will be covered in chapter 2.

UNDERSTANDING THE HEAVENS

The first thing we notice about the stars—other than their great number and beauty—is either their brightnesses or their patterns.

Let's consider brightness first.

Brightness and Magnitude

The ancient Greek astronomer Hipparchus divided the stars into six classes of brightness, called *magnitudes*. Stars of the first magnitude were the brightest, those of second magnitude the second brightest, and so on to stars of sixth magnitude, the dimmest ones visible to the naked eye.

In modern times, telescopes have made tremendously dimmer stars visible so that it has become possible to talk of seeing 10th-magnitude, 15th-magnitude, and even fainter stars (the higher the magnitude number, remember, the fainter the star). Some of the brightest stars actually deserve a class even brighter than first magnitude; these are considered 0 magnitude or even have a negative-number magnitude.

Astronomers have become capable of measuring the brightness of stars more accurately, requiring a more precise magnitude scale. Since a typical star of the traditional 1st-magnitude class was about one hundred times brighter than a typical star of the traditional 6th-magnitude class, astronomers decided that a difference of five magnitudes (the difference between magnitude 1 and 6 is five) would be a difference of one hundred times. (A slightly awkward consequence of this is that a difference of one magnitude has to equal 2.512—actually 2.5118 . . . times—because this number multiplied by itself five times equals 100.) Astronomers also started using decimals to express brightnesses more precisely. Thus a star midway in brightness between ones of 1.0 and 2.0 would be a star of magnitude 1.5 (a half-magnitude dimmer than the 1.0 star and half-magnitude brighter than the 2.0 star).

On the modern magnitude scale, the apparent brightness of the brightest star, Sirius, is –1.5. The brightest planet, Venus, can shine as bright as –4.7 sometimes. The Full Moon is –12.7. The Sun is –26.7. On the other hand, the faintest stars seen directly through large telescopes by the absolutely keenest-eyed observers under superb sky conditions shine at about magnitude 20. And with its long exposures and electronic imaging, the Hubble Space Telescope has recorded objects as dim as about magnitude 30.

The Constellations

The stars come in many different brightnesses. That is immediately obvious as you look up into a clear, really dark night sky. What is also apparent is that there seem to be various interesting patterns of stars. Not content to make mere triangles and other geometric figures in the patterns, ancient people saw outlines of animals, heroes, and many other things and named them as such: Leo the Lion, Orion the Hunter, Sagittarius the Archer. Those patterns and the people, beasts, and other things that they are supposed to represent have become accepted throughout the world as the official star patterns—the constellations.

The word *constellation* means a "togetherness of stars." Many of the constellations we now recognize seem to have been originally imagined by Mesopotamian cultures but received their final form in the hands (or rather, eyes and minds) of the ancient Greek and Roman astronomers and compilers. Forty-eight constellations recorded almost two thousand years ago by Ptolemy have more or less endured to the present. Twelve of the forty-eight are especially famous, even though some of the twelve are quite faint. They are famous because they are the constellations through which the Sun, Moon, and planets travel (even though we can't normally see stars in the daytime sky, we can still figure out which constellation the Sun is currently in). These twelve constellations form a band which stretches all the way around the heavens. Collectively, they are known as the *zodiac*. The zodiac constellations are, in their traditional order: Aries the Ram, Taurus the Bull, Gemini the Twins, Cancer the Crab, Leo the Lion, Virgo the Virgin, Libra the Scales, Scorpius the Scorpion, Sagittarius the Archer, Capricornus the Sea-Goat, Aquarius the Water-Bearer, and Pisces the Fish. We will meet with all these key constellations in this book.

Forty-eight constellations seemed to be enough for ancient Greek and Roman skywatchers. But in modern times there have been many others invented and added to the list.

The primary reason why additional constellations had to be invented was to organize the stars of Southern Hemisphere skies, which were too far south for the ancient Greeks and Romans to see. The first Europeans to witness these stars were explorers of the fifteenth and sixteenth centuries. Most people today know that there are some stars and star patterns which can't be seen at some northerly latitudes, like those of New York, Chicago, San Francisco, and London. The Southern Cross is the most famous of these star patterns. Most of the dimmer far-south constellations were created by just a few astronomers in the fifteenth through eighteenth centuries and were pictured as representing tropical creatures unknown to the ancients and various tools and other devices.

There were also regions of dim, unused stars between even the more northerly constellations known since classical times, so astronomers formed new constellations in these areas of the sky, too.

The final major development in the establishment of the constellations came in 1930. That's when the International Astronomical Union agreed on the precise boundary lines which would demarcate those areas of sky that belonged to each of the eighty-eight official constellations.

The Names and Designations of Stars

The existence of constellations has helped astronomers provide handy designations for thousands of stars. A few dozen stars are today still often called by proper names, names which have arisen for them over the course of hundreds or, in a very few cases, thousands of years. Names like Sirius, Betelgeuse, Regulus, and many more (including Polaris for the North Star) are well known. A few hundred more stars have proper names which are sometimes used by astronomy enthusiasts. But what about the other few thousand stars readily visible to the naked eye in a dark, clear sky? Many of these have received Bayer designations, also known as Greek-letter designations, based on the constellations they are in.

The basic idea of this system at first seems to be to name the brightest star in a constellation by the first letter of the Greek alphabet, followed by the (usually Latin) genitive form of the constellation's name. Thus, the brightest star in Leo the Lion, which has the proper name Regulus, is designated Alpha Leonis—the "first (in brightness) belonging to Leo." The second brightest star in a constellation would then be given the second letter of the Greek alphabet—Beta Leonis—and so on down the alphabet. Unfortunately, in many constellations the use of the letters was not done according to brightness but according to other criteria like location in the constellation or even rather haphazardly, as far as anyone today can figure out. Nevertheless, the Greek-letter system provides designations for many stars, and in most constellations you can count on the Alpha star being the brightest or at least one of the brightest in the constellation.

As astronomers saw more and more stars with better and better telescopes, they needed thousands of additional designations. The Flamsteed numbers are given in order of increasing easterliness in each constellation and there are some naked-eye stars of great interest with this designation—for instance, 61 Cygni, the sixty-first Flamsteed star in Cygnus the Swan. Ultimately, as the number of stars recorded has grown, astronomers have had to provide them with numbers from larger and larger catalogs to serve for identification purposes.

Right Ascension and Declination on the Celestial Sphere

How do we define exactly where a star is in the sky, even if we do have a name or designation for it? How do we describe an exclusive "address" in the heavens for each star?

The answer is by the equivalents of latitude and longitude in the sky. The effective equivalent of latitude in the heavens is called *decli-*

nation. The effective equivalent of longitude is called *right ascension,* or *R.A.* for short.

To apply these positional coordinates to the heavens, we must imagine that the combined heavens above and below our horizon are the inside surface of a sphere surrounding the Earth. This *celestial sphere* corresponds in all of its points to those of the Earth below. For instance, straight overhead for an observer at Earth's North Pole is the *north celestial pole.* The north celestial pole is the spot in the sky which remains fixed (doesn't seem to move at all) each day and night as the Earth spins. The reason for this is simple: the north end of Earth's rotation axis points straight at that spot in the heavens. There is likewise a south celestial pole which is located right overhead for an observer at Earth's South Pole. The only difference between the two celestial poles is that in our era of history there is no bright star near the south celestial pole. Second-magnitude star Polaris is called the North Star because it lies very close (though not quite exactly at) the north celestial pole.

There is also a *celestial equator* which, for an observer at Earth's equator, passes right overhead. Just as the Earth's equator is said to be at 0° latitude, the celestial equator is said to be at 0° declination. There is only one difference between how we designate other latitudes on Earth and their corresponding declinations in the sky. Latitude is measured in degrees North or South of the equator, up to 90° N(orth) at the North Pole and 90° S(outh) at the South Pole. But declination is measured in degrees + or – from the celestial equator, up to +90° at the north celestial pole and –90° at the south celestial pole.

Longitude and right ascension are a little trickier. For latitude and declination it is logical to begin the count of degrees at the equator and celestial equator, respectively. But in longtiude and R.A. the selection of where to draw your beginning line or meridian is a little more arbitrary. On Earth, the prime meridian for longitude is Greenwich, England, whose observatory is where many of the measurements and calculations concerning longitude were first made. On the celestial sphere, the prime meridian is drawn north-south through the vernal equinox or March equinox point of the heavens. Suffice it to say for now that this March equinox is the point in the heavens where the Sun is located when it is crossing north over the celestial equator, marking the start of spring in Earth's Northern Hemisphere and autumn in its Southern Hemisphere. Longitude on Earth is measured

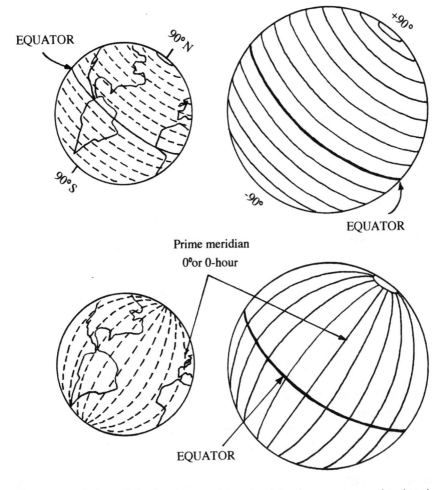

Figure 1.1. R.A. and declination on the celestial sphere compared to longitude and latitude on Earth.

180° West and 180° East from the Greenwich meridian. It is more useful, however, to measure R.A. continuously eastward from the meridian of the March equinox in 24 hours of right ascension rather than degrees. Thus the famous Pleiades star cluster is located at a little over 3h (3 hours) of right ascension, the star Fomalhaut at almost 23h of right ascension.

Like degrees of latitude and longitude, the degrees of declination and right ascension need to be subdivided into smaller units for precise usage. Each hour of R.A. is divided into 60 *arc-minutes of R.A.* (symbolized by *m*) and each arc-minute of R.A. is divided into 60 *arc-seconds of R.A.* (symbolized by *s*). Each degree of declination is divided into 60 *arc-minutes of declination* (symbolized by ') and each arc-minute of declination is divided into 60 *arc-seconds of declination* (symbolized by "). Thus when I look up the celestial coordinates of the bright star Vega right now, I find that this star is at a right ascension of 18h36m56.2s and +38°47'01". Actually, celestial coordinates change slowly due to a motion of the Earth called precession that we will discuss a little later in this chapter. For this reason, we do need to specify a particular point in time—for instance, the beginning of the year 2000—when the coordinates given are exactly correct.

The Seasonal Progression of the Constellations

There are winter constellations, summer constellations, and other different constellations at different times of year. This progression of constellations is the basis for the entire month-by-month organization of this book: the stars you see on December evenings are not the same ones you see on June evenings.

But what is the cause of this progression? It cannot be a true motion of the stars because they are so incredibly far away that any motion of theirs appears very slight in the sky over the years and even over the centuries. No, the movement which produces seasonal progression of the stars is the orbital motion of the Earth around the Sun, which keeps bringing constellations into different positional relations with the Sun. For instance, in the middle of a January night, we see the constellation Gemini at its highest in our sky. Also in January the constellation Sagittarius is at its highest around noon—and therefore is hidden by the glare of the Sun. In the middle of a July night, when Earth is on the opposite side of its orbit, the situation is reversed: we are facing Sagittarius at its highest around midnight and Gemini is at its highest around noon and hidden by the glare of day.

What we see is an evening constellation keep setting sooner and sooner after the Sun as the constellation's season of visibility is passing. Finally, that constellation begins setting so soon after the Sun that it is lost in the evening twilight low in the western sky. If you get up before dawn a few months after, however, you will see the same constellation in the east—and can thereafter watch it rising earlier and earlier before the Sun rises. How much sooner (by clock time) does a star rise or set each night? About 4 minutes. That translates to about 4 x 30 minutes = 120 minutes or 2 hours each month. And 12 x 2 hours = 24 hours every year—bringing the star back to its original apparent position after one year—one sidereal year. Sidereal means "of the stars" and sidereal time corresponds to the hour of right ascension which is on the north-to-south meridian of the sky. That is, if the right ascension on the meridian is 5h (5 hours), the time is 5h sidereal time. Sidereal time shifts in relation to the solar and clock times which govern our lives, so the 5h line of right ascension is on the meridian around midnight in December but around 4 A.M. in October and around noon in June. It is 5h sidereal time at all of those different clock times in different months. And the cause of this shift in the relation is the Earth's orbital motion around the Sun. This is what brings forth the constellations in their appointed seasons.

Circumpolar Stars

Having discussed R.A. and declination, we are better able to understand a famous behavior of some of the stars. I refer to the fact that those of us in the Northern Hemisphere see stars in the north sky that always stay above the horizon, and those of us in the Southern Hemisphere see stars in the south sky that always stay above the horizon. Such stars never rise or set. They travel in circles around a celestial pole in our sky but never go below the horizon. They are called *circumpolar stars*. For much of the world's population—everyone who lives at 40° N or farther north—the Big Dipper is the most famous star pattern that is circumpolar.

At first thought, it seems strange that some stars should never rise or set. But if we consider the view of the celestial sphere that is possible from different parts of the Earth, this state of affairs begins to make sense.

Imagine you are an observer at Earth's north pole. You have the north celestial pole directly overhead. The North Star, very close to the north celestial pole, hardly moves at all. But if you watched it carefully for hours you would see that it took about twenty-four hours to travel a tiny circle around the celestial pole. The farther from the celestial pole a star is, the larger the circle that it has to travel. From your vantage point at Earth's north pole, the stars of the celestial equator

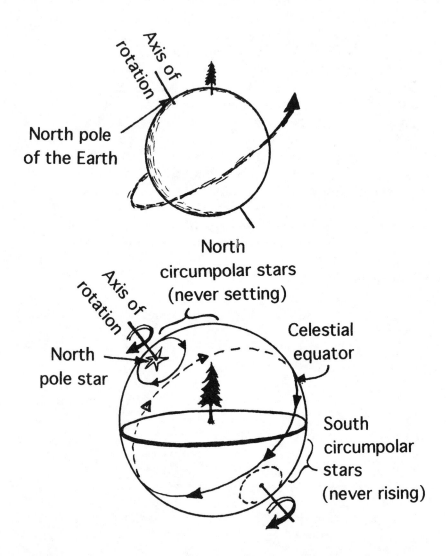

Figure 1.2. Real (top) and apparent (bottom) situations: Earth rotating east-ward (top); celestial sphere rotating westward around motionless Earth (bottom). Note how an observer in the Northern Hemisphere sees stars near the north end of the axis of rotation never go below the horizon ("north circumpolar stars") and stars near the south end of the axis of rota-tion never come up above the horizon ("south circumpolar stars").

would be right on the horizon, traveling rightward on that the largest circle of all. As an observer at the North Pole you would see that all stars above your horizon would always be circumpolar. But these stars of the north celestial hemisphere would be the only ones you would ever see. Half of the heavens—the south celestial hemisphere—would be forever hidden from your view.

The situation would be very different for an observer on Earth's equator. There, the north celestial pole would always lie right on the due-north point of the horizon, and the south celestial pole would always lie right on the due-south point of the horizon. No stars would be circumpolar. All stars would rise and set. And no part of the heavens—neither the northerly Big Dipper nor the southerly Southern Cross—would be hidden all the time.

Of course, most of us live somewhere between Earth's poles and equator. The majority of Earth's population lives in the middle latitudes, mostly in the Northern Hemisphere. Let's consider the situation at such a location—let's take halfway between the pole (90°N) and equator (0°), at 45° N latitude.

At 45° N, we would find the North Star and north celestial pole halfway up the sky from the due-north point on the horizon. All stars within 45° of declination from the celestial pole (those between +45° and +90° declination) would be circumpolar, pursuing circles around the pole in the north sky. On the other hand, all stars within 45° of declination of the south celestial pole (those between −45° and −90° declination) would be hidden from view all the time. The rest of the stars—those that are neither north circumpolar nor so far south as to be always below the horizon—would be those between +45° and −45° declination. And those stars would be the ones rising in the east (more precisely, some would rise north of due east, others south of due east, etc.), progressing across the sky, and setting in the west. If you stopped to think about it, you would realize that even these stars were pursuing circles around a celestial pole—but circles so large that the east and west horizons cut them off into arcs.

After all this talk of stars making circles, rising and setting, it is good to remind ourselves that these motions of the stars are just *apparent* motions. They are all really the result of Earth's rotation. As Guy Ottewell, who has provided the maps for this book, has so beautifully put it: "Horizons roll, and not the standing stars."

Precession

There is a complication in keeping track of the positions of stars, even in keeping track of which stars are circumpolar, over the course of time.

The right ascension and declination I gave for Vega a few pages back was actually only strictly correct for "epoch 2000"—for precisely at the start of the year 2000. The most important reason we must specify an epoch date is not because of change in position caused by the true motion of stars through space. The stars are so distant that this true motion translates into a very tiny proper motion—that is, change in position in the heavens caused by motion through space. The proper motion of stars in our heavens is tiny for all but two kinds of stars: (1) the closest stars and (2) slightly more distant stars which are moving unusually fast through space and/or at nearly right angles to our line of sight with them. No, the far larger change in a star's R.A. and declination as mere years and decades pass is caused by something else: a movement of Earth's axis called *precession*.

Picture the Earth spinning like a child's toy, a top. The gravitational pull of the other planets introduces into the spin of this giant top a wobble of extremely long period. The rotational axis remains tilted at about 23 1/2° from the plane of Earth's orbit around the Sun. But the direction in space to which the axis points changes very slowly, describing a circle 23 1/2° in radius among the stars. In other words, the position of the north celestial pole at which the north end of Earth's axis points traces out this circle against the background of stars. How long do the celestial poles take to complete this large circle in the heavens? The answer is 25,800 years! That's a long time in human terms but the rate of precession is still large enough to make a change in a star's celestial coordinates that is significant in just a few years for the purposes of amateur astronomers, let alone professional astronomers. So astronomers specify the epoch year for which R.A. and declination are given.

Star atlases and catalogs which used to be based on epoch 1950 coordinates are now given in epoch 2000 coordinates, and eventually we will see epoch 2050 coordinates. As a matter of fact, as amateur astronomers have gained more sophisticated instruments and undertaken observations requiring greater precision, it has often become necessary for them to specify celestial coordinates closer to the precise time of observation.

Since precession changes the position of the celestial poles in the heavens, Earth gets a variety of North Stars (and South Stars) over the course of the 25,800-year period. Likewise, "precession of the equinoxes" changes which stars are in the day sky and night sky at the equinoxes—start of spring and start of autumn—and thus which stars and constellations we see on winter, spring, summer, and autumn evenings. If precession were a great amount faster, a book like this one, based on what you can see in the starry sky each month of the year, would have to be updated before many years had passed. Since precession is slow, however, our descriptions of which stars are visible each month should be valid for centuries.

Altitude and Azimuth

The equatorial system of celestial coordinates—the one which utilizes the celestial sphere and R.A. and declination—is the most useful one for astronomical observers, specifying the position of a star independent of time of observation. But another system is sometimes useful, too, if we do specify the time of the observation.

The altazimuth system deals with not the celestial sphere but the half-sphere or seeming dome of the sky. Here our coordinates are altitude and azimuth. Altitude in this system is an object's height in the sky—not in miles or thousands of feet but rather in degrees of angular measure. The sky is a half-sphere and therefore measures 180° from one horizon through the overhead point to the opposite horizon. Altitude can consequently run anywhere from 0° at the horizon (the meeting point of the sky and land or sea) to 90° at the overhead point, or as we more technically call it, the *zenith*. Altitude is vertical measurement in the sky. Azimuth is horizontal measurement in the sky. We could give it in cardinal directions like north, southeast, or, more precisely, south-southwest. But customarily astronomers give it in even more precise measure, in 360 degrees of azimuth around the sky. In this system, due north is the starting point, with an azimuth of 0°. Due east is 90°, due south 180°, due west 270°, and so on back to 0° to complete the 360° circle.

The altitude and azimuth of a star keeps changing so we must specify the time of observation—unless it is a star very near a celestial pole. If we don't bother to be very precise (Polaris is slightly off from the north celestial pole), we can say that Polaris stands always at the same azimuth and altitude for an observer at a given location. The azimuth is 0° wherever your location. But as we noted in our discussion of circumpolar stars, the altitude of Polaris in your sky will depend on your latitude. The handy rule of thumb is that whatever your latitude is, that is the altitude of Polaris in your north sky. Thus if you live at 40° N, Polaris will appear at an altitude of 40°.

By the way, if you know your latitude or note the altitude of Polaris, you can figure out what is the declination of objects which pass through your zenith and what is the limiting negative—that is, southern—declination beyond which stars never rise into view at your location. If we know that Polaris is 40° altitude and that the distance to the zenith is another 50°, we can just subtract 50° from the— roughly—90° declination of Polaris and determine that objects which pass overhead at our site should be those at +90° – +50° = +40° declination. Since it is 90° from the zenith down to the south horizon where the most southerly stars we can see peek into view briefly, then +40° (declination of objects at our zenith) –90 degrees = –50° declination. An observer who lives at about 40° N—say in Philadelphia, Denver, or San Francisco—can see stars as far south as –50° declination.

Angular Measure and Universal Time

Suppose you don't know your current latitude. Maybe you are on a cruise ship. If you are in the Northern Hemisphere and can measure the angular altitude of Polaris, you can determine your latitude. But how can one easily measure angular altitude and other angular distances in the sky?

A literally handy way to do this is by knowing the angular width of your hand when you see it out at arm's length. For almost everyone, this width is approximately 10°. You might think the figure would be different for large adults versus small children. But even though a tall basketball player's fist may be much wider than a young child's, the player will also have very much longer arms. The big fist will be seen from farther away. So if you are trying to figure out whether you really are at, say, 20° N latitude, see if you can fit two widths of your fist between Polaris and the horizon.

Using your hand for angular measure will more often be valuable for things like trying to point out a particular star to someone. You can say that the star you want them to see is "about 40° (four widths of your fist at arm's length) above the top of that pine tree" or "15° to the right of that really bright star." You will find many other uses for this means of angular measurement. And for smaller distances, try the width of your fingers or thumb at arm's length. Your thumb may be about 3° wide. Perhaps that is the distance in the sky between two planets tonight.

For telescopic observations we need even smaller units of angular measure. So we resort not just to degrees but also minutes (') and seconds of arc (")—1° = 60' and 1' = 60". The Sun and Moon appear 1/2° or about 30' wide. Jupiter averages about 40" wide and distant Neptune appears only about 2" wide as seen from Earth.

Astronomers measure distances in the sky by angles. But how do they measure time?

The answer, often, is by UT or Universal Time. Universal Time is twenty-four-hour time whose starting point of each day is that of the Greenwich meridian, the 0° longitude line on Earth. Thus 12 A.M. midnight standard time in Great Britain—through which runs the Greenwich meridian—is also the start of a date in Universal Time. It would be expressed as "0 UT". 1:30 P.M. standard time in Great Britain would be 13:30 UT.

The time of astronomical events is often given in UT, so you must know how to convert UT to your local time and vice versa. In the United States, Eastern Standard Time (EST) is 5 hours behind UT—in other words, an event at 9 UT on April 3 would occur at 4 A.M. EST on April 3. Central Standard Time (CST) is 6 hours behind UT, and so on. Note that an event occuring at 3 UT would take place at 10 P.M. EST of the previous date. (A helpful reminder: the start of a new UT day occurs at 7 P.M. EST and 8 P.M. EDT (Eastern Daylight Time), 6 p.m. CST and 7 P.M. CDT, etc.)

Meteors and Auroras

There are different constellations visible in the evening sky each month. But there are a few other sky sights much closer to us than the stars which are also seen more favorably in certain months of the year. They are therefore mentioned in the chapters for those months in this book.

One of these sights is the usually fraction-of-a-second streaks of light popularly called falling stars or shooting stars. The scientific name for these phenomena is *meteors*. They are actually caused when pieces of space dust and rock from the dirty ice balls that are comets and rocky worldlets that are asteroids enter our atmosphere at tremendous speeds (many in excess of 100,000 mph) and burn up from friction, usually when fifty or more miles above the Earth's surface.

For reasons unknown, the second half of the year tends to offer random meteors in greater numbers—up to maybe ten per hour in the best sky conditions. It also offers most of the year's major *meteor showers*. A meteor shower is an enhanced number of meteors all appearing to diverge from a single spot among the constellations called a *radiant*. Meteor showers occur on approximately the same dates each year. Some of their meteors may leave glowing ionization trails after the meteors themselves have vanished. Some of the meteors may be *bolides*, meteors that explode. A few of the meteors may be *fireballs*, meteors brighter than Venus or Jupiter (on rare occasions, a fireball may rival or even surpass the Moon in brightness)!

An observer should keep a count of how many meteors he or she sees each hour and/or in shorter periods (fifteen minutes, thirty minutes, etc.). Don't combine the totals of meteors which you and fellow observers see, and don't count in your official tally a meteor you wouldn't have glimpsed if another observer hadn't shouted out. Try to determine the *limiting magnitude* (the magnitude of the dimmest stars you can see at the time of observation) overhead or in the part of the sky where you are looking. Where should you look? In a meteor shower, meteors coming almost directly at you can appear as short streaks or even brief points of light near the radiant. But meteors in a shower can first become luminous anywhere in the sky—it is just that if you mentally extend the visible path of the meteor backward from where it first appeared, your line would eventually bring you back to the radiant. The best place to look to see the most meteors is probably a moderate angular distance—say 30° to 50°—away from the radiant.

The best annual meteor showers can serve up as many as sixty or more meteors per hour around the time of their maximum on their best night. On very rare occasions, however, rates of more than one thousand meteors per hour may be seen and this awesome spectacle is called a *meteor storm*.

There is another literally moving sky phenomenon which tends to occur most and best at certain times of the year. It may appear as a glow which can fluctuate slowly or rapidly in brightness throughout its various parts. Within that glow this phenomenon can exhibit different colors and can form rays, arches, curtains, and other structures which may move, sometimes very rapidly. The phenomenon is called the aurora borealis or Northern Lights. The name is given with good reason, for vivid displays of it are fairly uncommon unless you live in a place like Alaska, Canada, the northernmost tier of the contiguous United States, Great Britain, or Scandinavia—in other words, at a fairly high northern latitude. Actually, the phenomenon can also be seen at far southern latitudes, though there is less land far enough south for as many people to see this aurora australis. Scientists usually call any display of the phenomenon by the more inclusive term *aurora*, leaving off the "borealis" or "australis."

What causes the aurora, the sky's greatest light show? A very quick and somewhat oversimplified answer: it is outbursts of atomic particles from flares and coronal mass ejections and coronal holes on the Sun which get accelerated in Earth's magnetic field and carried by our planet's magnetic lines of force to polar regions where they strike gases of our upper atmosphere and cause the gases to glow.

The aurora is most common for a few years around what is called *solar maximum*—the peak in solar activity like sunspots and flares—which occurs about once every eleven years. (The last solar maximum was in the year 2000, the next may be in 2011.) But the aurora is also most common around the equinoxes, the starts of spring and fall. This is at least partly because those are the times of year when the active latitudes of the Sun are pointed more directly toward the Earth.

The Milky Way

We move from the electrifyingly lively moments and movements of an aurora or a strong meteor shower to a sight of awesomely dreamlike stillness. Yet it is a sight which, like meteors and auroras, is best seen at particular times each year.

Astronomers can mean one of two related things when they speak of the Milky Way. As we'll see in the next chapter, they use this term to refer to the enormous system of many billions of stars in which our own Sun and solar system exist. But for observers the Milky Way is a band of softly glowing light which can be seen arching across the sky. For evening observers, a dim section is visible in winter, a more peripherally located but fairly prominent section in fall, and a brightest and most impressively structured section in summer.

Under superb sky conditions, the Milky Way band shows intricate structure, its glow pooled here and there into brighter star clouds and cut into in other places by tongues of darkness, including one largest dividing split called *the Rift*. Even before you find out what causes this delicate but magnificent phenomenon, it inspires awe and contemplation. Prescientific cultures have thought the Milky Way band was a path for the souls of the dead or the eerie spine of the heavens, the "backbone of night."

USING YOUR SKIES AND EYES

Everyone knows that a cloudy sky prevents us from seeing celestial sights. But most beginners don't realize that our view of the stars can be severely compromised by three other things. One is haze, which causes a dramatic dimming effect, and is often present even when there are no clouds in the sky. The second is bright moonlight. The third is light pollution. We'll start with the most troublesome of these three, but one that is capable of being reduced by human action— light pollution.

Light Pollution

If you live in or anywhere near a medium-size city today the glare of city light pollution will limit your naked-eye observations to seeing just the Moon and planets (if they are up) and only the brighter stars. This is somewhat like what people used to see on nights when the sky was very hazy or on the night each month when the Full Moon was washing out most celestial sights with its overwhelming glow.

Just think how much more the night sky should be able to offer us. How much more visually splendid are the sights on a clear night at a rural location many miles from city lights! From such a location you don't have to start out making a deliberate effort to seek those sights as you do in the city. The sights seem to seek you, and there are very many more of them in both number and kind. The stars burn astonishingly brighter and crowd the sky by the thousands. The constellations all flame forth in their ancient glory and story. The shining streaks of meteors frequently punctuate the night. Furthermore, at some part of the night, the sight which is not the brightest but is arguably the grandest vision of all, the dreamlike glow of the Milky Way, arches across the sky in its majesty. Yet all these sights are lost for up to tens of miles around a city with wasteful lighting, whose light pollution is severe.

Light pollution may be defined as excessive or misdirected artificial outdoor lighting. It is by far the greatest threat to the hobby of astronomy and to all enjoyment of the night sky. Loss of our direct connection to the universe would be a crushing blow to the human spirit. But there are also several other serious negative effects of light pollution. Some of these involve money and saftey, matters which usually impress politicians and the public more than hearing about astronomy.

Skyglow is the component of light pollution which is useless illumination of the atmosphere above cities. It wastes billions of dollars annually and the energy-equivalent of millions of barrels of oil a year, even just in the U.S. alone. The component of light pollution called *glare* (also largely due to poorly shielded lights) shines into the eyes of motorists, reducing traffic safety, and into the eyes of homeowners and law-enforcement officials, thwarting crime prevention.

What can you do about light pollution? A lot. A first step is going to www.darksky.org and getting information from IDA, the International Dark-Sky Association. At the time of this writing, IDA has almost ten thousand members in about seventy countries around the world. You should be one of them!

Phases and Hours of the Moon

The light of a thin crescent Moon has little significant effect on observations of stars in the rest of the sky. But when the Moon is at a larger phase, its radiance will reduce the darkness of the entire sky and therefore the visibility of faint stars and other deep-sky objects.

Fortunately, there are fairly simple rules for determining when the Moon will be in the sky.

For a few days after New Moon, the Moon is a thin crescent setting only an hour or two after the Sun. It is not bright enough to interfere with any observations of the stars. With each successive night, however, the Moon sets very roughly an hour later and its lit part gets noticeably thicker. By the time the Moon sets around the middle of

Figure 1.3. Satellite image of the United States at night in 1979. (Courtesy of International Dark-Sky Association.)

the night and is highest in the south at nightfall, it appears half-lit and is quite a hindrance to observing faint parts of the starry heavens. Despite being a half-Moon, this phase is called First Quarter—because the Moon has completed the first quarter of its month-long journey from New Moon through all its phases back to New Moon.

It takes about a week to go from New Moon to First Quarter Moon. During the next week, the Moon goes from First Quarter to Full Moon. In this period, the Moon keeps getting brighter and setting later and later in the night. The dark hours of the night become fewer and fewer. In fact, Full Moon rises around sunset, is highest in the middle of the night, and sets around sunrise—so this brightest Moon is visible all night long and many kinds of observations of the stars are severely compromised all night long.

During the week from Full Moon to Last Quarter, however, the situation improves for those who want to observe the starrry heavens. The Moon is rising about an hour later in the evening each night and its phase is getting less fat and bright. More and more of the evening hours become available to observing the stars in moonfree skies. Last Quarter Moon doesn't rise until the middle of the night and is at its highest in the south at dawn.

After Last Quarter, the Moon is a waning (shrinking) crescent coming up sooner and sooner before dawn. So most of the night is free of moonlight and when the Moon does rise it doesn't effect star observations too severely. Finally, the Moon has its dark side to us again and rises and sets with the Sun and is at the phase called New Moon—when it poses no problems whatsoever for viewers of the stars.

To summarize then: *between New Moon and First Quarter*, the Moon keeps setting later in the evening and becoming brighter and therefore more of a hindrance to observing the stars; *between First Quarter and Full Moon*, the Moon is up and more and more of a problem until later and later in the night and at Full Moon is at its brightest and spoiling our view of the starry sky all night; *between Full moon and Last Quarter*, the Moon keeps rising later in the evening so that more and more of the evening is free of moonlight for star-watchers; finally, *between Last Quarter and New Moon*, the Moon doesn't rise until the hours after midnight and becomes thinner and thinner and less of a hindrance to star observations as it rises sonner and sooner before dawn. For a few days around New Moon, the Moon is rising and setting near the Sun and is not seen at night—the perfect situation for those who are concentrating on observing the wonders of the starry heavens.

An Aside about Planets and Stars

Even people who never deliberately look up at the night sky will sometimes have their gaze irresistibly drawn up to a dramatic sight: a brilliant point of light hanging right beside the lunar crescent. A close meeting of two heavenly objects—more precisely, the moment when they have either the same right ascension or *ecliptic longitude* (another west-east measure of position)—is called a *conjunction*. A conjunction can occur between the Moon and a planet, the Moon and a star, a planet and a star, or two planets. (If the Moon passes right in front of a planet or star, or a planet passes in front of a star, the event is called an *occultation*.) But here is our key question: is the point of light near the Moon on this night a planet or a star?

Well, even a very thin crescent Moon is still quite bright. So any object which stands out very prominently near it must be bright, too, and is most likely to be a planet, in fact one of the brightest planets.

Mere brightness is not in itself a perfect test of planethood, however. Three planets are so distant and dim that it was not until the past few hundred years that they were discovered—with telescopes. More importantly, even a few of the five planets bright enough to be known to naked-eye observers throughout history are sometimes outshined by a number of the brightest stars.

Fortunately, there is another instant test of the nature of a bright point of light that you can use if you don't have a telescope to magnify and look for the globe of the planet. You can see if the point of light twinkles.

Since ancient times, people have noticed that planets almost always shine with a steady (or nearly steady) light, while stars twinkle.

Stars are so distant that even in big telescopes they still look like points of light. Light coming from this point source is shifted around a bit as it follows a path through the ceaseless turbulence of Earth's atmosphere. But, as telescopes reveal, when planets are magnified their globes are shown to have some appreciable apparent size. If light coming from one edge of the globe of Jupiter is shifted away from us by a cell of turbulence up in our atmosphere, light coming from the rest of the disk may still reach us fairly straight on. The overall effect is for the naked-eye planet to appear to us as a steady light, twinkling very little most of the time. (There are exceptions, rare times when a planet may twinkle somewhat. For instance, when very low in the sky a planet will twinkle because it then has to shine through a much longer pathway of air, the grazing path along the curve of the Earth.)

There is a final habit of planets which can distinguish them from stars if they are observed over a number of nights: planets change their positions relative to the unchanging patterns of the background stars. As a matter of fact, the ancient Greek word from which we get the English word *planet* means "wanderer."

One night, a planet might appear to form a perfect line with the two bright stars of Gemini the Twins, Castor and Pollux. The next night, the line might appear slightly crooked. It is the planet that has changed its position relative to the two stars. In a week (for a fast planet) or year (for a slow planet) the "wanderer" would leave the two stars far behind.

The fascination of the planets is great but this is a book on the stars. So we will conclude this chapter by discussing how to judge when hazy skies will hamper studying the stars and how to use special observational techniques to see the stars better.

Getting the Most from Eyes and Skies

You will need to be far from city lights and have a clear, moonless sky to see the Milky Way properly and to see many of the dimmer heavenly sights at all. But it also helps to know how to use your eyes in special ways. This chapter has been confined largely to what we can see with the naked eye, without optical aid. But your eyes are themselves optical instruments. It is appropriate that we end this long chapter with a look at looking—and also how to find the clearest and darkest sky to look into.

The human eye utilizes two separate types of vision for bright and dark conditions. The pupil of the eye opens wider in darkness, of course, but there is actually a different type of cell in the retina which is needed for low-light situations and a chemistry of the eye which takes a certain amount of time to make the eyes fully sensitive to dim sources of light in darkness.

When you step out under a reasonably dark sky, with no nearby light shining in your eyes, your eyes begin their *dark adaptation*. With each passing minute you will notice that you can see fainter stars, and for about the first twenty minutes the improvement is quite dramatic. A lesser improvement continues for a long time. But if you should be exposed to even moderately bright light then the process will be undone and you will have to wait for minutes again to regain proper dark adaptation. One of the cardinal sins at a gathering of amateur astronomers is to shine a flashlight up into someone's view. Interestingly, however, dark adaptation is not spoiled if the light is red. Thus many amateur astronomers purchase flashlights with red filters or simply tape red cellophane over the bulb end of an ordinary flashlight.

The other most important help to seeing dim stars is using *averted vision*. The central region of our retinas is rich in *cone* cells, which register color and provide us with sharp vision. But farther out away from the center of our vision it is the *rod* cells that dominate. Rods don't show color or provide sharp imaging (outside of a surprisingly small central area of our view where the cones dominate, our vision is blurry). But the rods are sensitive to low light levels. Therefore, to see fainter stars, try directing your gaze slightly away from the object you are trying to see. You will be bringing its light onto the part of your retina with the most rods and you will find that the dim object pops right into prominence.

Of course, you can't see very faint stars if your sky's background is too bright from light pollution. But even light pollution's effect can be somewhat minimized if you choose the clearest nights. Humidity (water vapor) and haze from dust and other particles in the atmosphere not only absorb the incoming light from stars, they also scatter starlight (plus any light pollution), leading to a bright, washed-out sky background. What we want is a night when the atmosphere has good *transparency*—that is, the quality of being able to pass the maximum amount of astronomical light through it with the minimum of scattering and absorption.

How can you tell long before sunset that a night of excellent transparency is in store? Apart from elements of the weather forecast, there are signs you can look for in the daytime sky. The best nights should follow days of deepest blue sky. But note especially how close to the horizon the sky is still blue and even more especially how close to the Sun you can still see blue. On days of good transparency the Sun will have only a small *aureole*—that is, a small region of whitish sky caused by scattering of light around the Sun. On days of truly great transparency (quite rare in most climates), there will be no aureole around the Sun.

Of course, periods of what we can call "severe clear" (excellent transparency) are apt not to last very long. The sky that looks so good at 3 P.M. might worsen in transparency by nightfall at 8 P.M. Or it might improve, bringing you the starriest night in your life. Each night is a new adventure. There is no telling what unexpected wonders may befall you. And no matter what happens, you will find peace and solace in your companionship with the heavens.

Chapter 2

Understanding and Observing the Universe

I f you've never yet learned a constellation or been able to identify which planet you are seeing in the sky, you may still have read the basics about the structure of the universe: planets circling around stars, stars circling within galaxies, galaxies fleeing from one another because of the expansion of the universe and enlargement of space itself since the big bang.

Within that basic outline, however, there are all the incredibly rich details that make for a great adventure. And to understand what you see through a telescope and thereby be able to enjoy the sights much more, you need to know these details. How do scientists measure the distances to the stars and figure out their true brightnesses? What do those true brightnesses when combined with the chemical spectra of starlight tell us about the nature of stars and the stages they pass through in their lives? Of more direct importance for the reader of this book who wants to observe these objects: what are the major kinds of stars, double stars, variable stars, nebulas, and galaxies—and what do they look like?

DISTANCE, LUMINOSITY, AND SPECTRAL CLASSES

The first property of the stars to recognize is the immensity of their distance from us. Since stars really are suns at least somewhat like our own, then clearly they must be enormously far away. It would take a tremendous distance for a blindingly bright radiance like our Sun's to be reduced to a gentle flicker in the night.

Light-years

In our solar system, we usually speak of distances in millions of miles or kilometers or at most a few billion of either. There is also the *astronomical unit* or a.u., which is the average distance from the Sun to the Earth, about 93 million miles or about 150 million kilometers. But to speak about the distances to the stars, we really need a new unit of distance—the *light-year*.

Light is the fastest thing in the universe, traveling about 186,000 miles per second in a vacuum. A light-year is the distance that light can travel in a year. I could tell you that a light-year is about 6 trillion (that is, 6 million million) miles or about 63,000 a.u. but famous astronomy and children's book author H. A. Rey made a much more helpful comparison. If you represent an a.u.—the Sun-to-Earth distance—by one inch, then a light-year is about one mile! On this scale, the average distance of Pluto from the Sun is about one meter— slightly more than one yard. Now the nearest star (other than our Sun) is more than 4 light-years away. So the difference between the distance to the farthest planet and the distance to even the nearest star is roughly the difference between a yard and four miles. Some of the most distant stars we can see with our naked eye are more than 4,000 light-years away and so on the scale we are talking about would lie more than four thousand miles from the Sun and its six-foot-wide realm of planets!

Distances and Luminosities

How have astronomers been able to determine the distances to the stars?

First of all, by *parallax*. Parallax is best demonstrated by holding your finger out at arm's length and looking at it with first just one eye open and then with just the opposite eye open. Do this quickly and repeatedly and you will notice a slight change in the apparent position of your finger against a much more distant background object—say, a wall or a lamp in another room. Now move your finger closer to your eye and repeat the experiment. The apparent change in position of your finger against the background is much greater. So the closer an object is to a viewer, the greater its parallax—the change in position due to a different viewing point—will be. But how, you may wonder, can we view objects as distant as the stars from vantage points that are sufficiently far apart? By observing a star at six-month intervals. At those two times our planet is on opposite sides of its orbit. Since the Earth's distance from the Sun is about 93 million miles, the diameter of its orbit is twice that value—and observing six months apart is like observing with first one eye and then the other one about 186 million miles away from the first. Even this baseline is not big enough to make any but relatively near stars change apparent position enough for the change to be accurately measured. Or at least that was the case until the Hipparcos satellite was able to make ultraprecise positional measurements of stars in the early 1990s. Hipparcos enabled us to reduce the uncertainty of star distances from 25 percent to 2 percent at 80 light-years, 45 percent to 8 percent at 300 light-years and from 50 percent to 25 percent even out at 1,200 light-years.

By the way, the distance at which the parallax of a star would be 1 arc-second is called a *parsec* and is equal to about 3.26 light-years. Therefore, even the nearest star is more than 1 parsec distant.

Other methods of distance determination depend on knowing things about the nature of certain stars. For instance, we have discovered that a type of variable star called a Cepheid has a true brightness which is related to its *period*—that is, the length of time it takes for the star to vary from one maximum in light output to the next. So if you observe a Cepheid's period and its apparent brightness, you can figure out its distance.

Many clues about the nature of stars have been learned by studying the spectra of their light. But before we discuss spectra and spectral types and classes, let's note the units which are used to measure the true brightness of stars once we know their distances accurately. One system of units is *absolute magnitude*. That is the magnitude that a star would appear to have if it was 10 parsecs from Earth—about 32.6 light-years. Our Sun would appear as a 5th-magnitude star, only dimly visible to the naked eye, if it was this far away.

On the other hand, the stars of greatest true (as opposed to apparent) brightness in our sky have absolute magnitudes of –8 or even brighter. If such a star was located 32.6 light-years from us, it would rival the brightness of a half-Moon in our sky. A measure of true brightness which I will use in this book far more often than absolute magnitude is *luminosity*. The luminosity of a star is its true brightness expressed in units of the Sun's true brightness. A few of the stars visible to the naked eye in our sky have a luminosity more than one hundred thousand times that of the Sun.

Spectral Types and Spectral Classes

In the nineteenth century, the invention of the spectroscope enabled astronomers to start determining which chemical elements were present in a star by studying the spectrum of a star's light. By passing starlight through a thin slit and then through prisms or a diffraction grating, a spectroscope can spread out the various wavelengths of a star's light somewhat like raindrops do to sunlight to produce a rainbow. The spectra of stars includes bright lines of emission (from hot gases) and dark lines of absorption (from cooler gases) which can be compared to those produced by light from chemicals in the laboratory. In this way, much can be learned about the chemical composition of stars even though they lie at vast distances from our world.

In the early twentieth century, the modern system of *spectral types* was instituted, with each type represented by a letter in a sequence ranging from hottest to coolest star. The sequence was: O, B, A, F, G, K, M (a famous mnemonic sentence for remembering this order is Oh Be A Fine Girl [or Guy], Kiss Me). A few additional special types have been added, such as N and C for certain cool stars and W for the very hot Wolf-Rayet stars. These spectral types correspond not only to surface temperatures but also roughly to colors: O through A stars are blue-white to white, F stars are white or yellow-white, G stars are yellow-white or yellow, K stars are orange, and M stars are a deeper orange tending toward red.

More precise specification of a star's spectral status can be made, and one way to classify the differences within each spectral type is by *spectral classes*. Within a given spectral type—say, G—the classes are numbered from 0 to 9, with 0 the hottest and 9 the coolest stars in the type. The famous yellow winter star Capella is a G8, indicating it probably has a cooler surface than the Sun, which is classified as a G2 star. If the Sun was a little hotter, it might be classified as a G1 or G0, or even be moved into the next, hotter spectral type, and be considered an F9 star.

The emission and absorption lines in the spectrum of a star can help tell astronomers many things. Among these are whether a star has a close companion sun and whether it rotates rapidly.

THE LIVES AND DEATHS OF STARS

What can knowledge of the above properties of stars tell us about how stars live and die? A great deal, as it turns out.

Red Giants and White Dwarfs

A graph which plots luminosity of stars on the vertical axis and spectral type (or surface temperature) on the horizontal is called a *Hertzsprung-Russell diagram* or *H-R diagram*. By figuring out why stars congregate in certain areas of the diagram, astronomers have been able to understand much about the nature of stars and piece together the different stages of the stars' lives.

For much of their lives, many stars lie along a "main sequence" which runs roughly from upper left to lower right on the diagram. Our Sun is a main sequence star. Such a star initially and most fundamentally produces its radiant energy by the fusion of hydrogen atoms into helium atoms. This is the thermonuclear source of the Sun's light and heat. But as such a star begins to use up the hydrogen in its core, the outward pressure of its radiation loses the battle against its own gravity. The core must contract to become hotter and be able to use its helium as thermonuclear fuel. The smaller, hotter core causes the outer layers of the star to expand enormously but cool and become extremely tenuous. And so the star has moved off the main sequence to become a *red giant*, vastly larger than our Sun currently is. The most famous red giant is the star Betelgeuse, which might be more than one thousand times wider than our Sun. But several billion years from now the Sun will become a red giant itself (although much smaller than Betelgeuse now is), when its thin outer layers will swell out perhaps as far as Earth's orbit, charring or even engulfing our world.

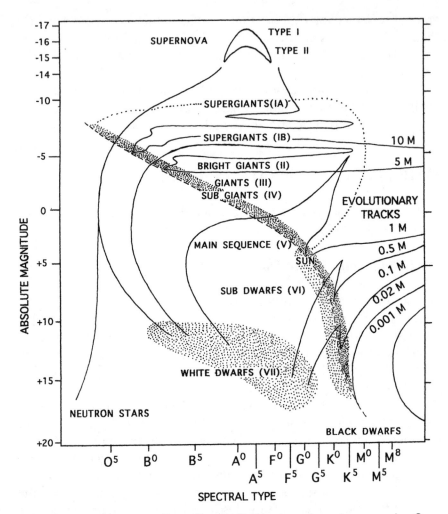

Figure 2.1. Hertzsprung-Russell or H-R Diagram, based on one by Guy Ottewell. The letter M on the middle right part of the diagram means "one solar mass" (thus "5M" is five times the mass, or amount of material, in our sun).

A star cannot remain as a red giant forever, though. It will eventually run out of helium and heavier elements to fuel it. If it has approximately the *mass* (amount of material) that the Sun does, it will suffer a gravitational collapse down to an object roughly the size of a planet such as the Earth. At that point it has collapsed down to an incredibly dense state of matter in which electrons have been stripped from atoms and mingle but still cushion atomic nuclei from each other. The remaining heat of the star compressed into so small a space makes the surface white and hot—it has become a white dwarf. That *white dwarf* will eventually fade to a dark stellar remnant but only over the course of many billions of years. (The most famous white dwarf is the much dimmer companion of Sirius, the brightest star in our sky.)

Red Dwarfs, Blue Giants, Neutron Stars, and Black Holes

What we have discussed so far is the life of a star whose mass is similar to that of our Sun.

A star much less massive than the Sun will burn weaker and cooler and live as a *red dwarf* a few times smaller than the Sun until it uses up its fuel and shrinks to a planet-sized white dwarf. (The most famous red dwarf is probably Proxima Centauri, a member of the three-star Alpha Centauri system, which is the closest star system to our own solar system.)

A star much more massive than the Sun has a more spectacular life and fate. It will burn its fuel quickly and live not for billions but only millions of years. It may be massive enough to burn as a *blue giant* (the most famous blue giant is Rigel, which with Betelgeuse shines in the mighty constellation Orion). When a massive star becomes a red giant, it evolves into a much larger one than our Sun will ever be. And its core is large enough to keep contracting and become hot enough to convert helium to heavier and heavier elements until it makes iron. Then the fate of the massive star is to become a *supernova*. In a supernova there is both a collapse (or implosion) and an explosion. A large percentage of the star's matter is either converted to energy or ejected out into space, sometimes into a glowing cloud that can be seen even in amateur telescopes on Earth for thousands of years—a supernova remnant or SNR.

But the collapse of the star's core in a supernova leaves that part of the star in one of two states.

If the star was only a few times more massive than the Sun to begin with, the core will shrink to a city-size object known as a neutron star. The neutron star is made of even denser matter than a white dwarf, for in it the subatomic particles are crushed together. It spins at speeds of as high as thousands of times per second, the rate decreasing over time. And if the neutron star has the right orientation toward Earth, we receive a burst of radio waves from each rotation of it and call it a *pulsar*.

If the star is many times more massive than the Sun to begin with, then not even the nuclear forces can resist its gravitational collapse and it becomes an object sometimes said to be "infinitely small" or to have no size at all: a *black hole*. Astrophysicists continue to debate the nature of black holes and generate new theories about them while observational astronomers identify more places where effects on surrounding stars and matter appear to indicate the presence of black holes. One important discovery is that at the center of many a galaxy there appears to be one or more black holes which have become supermassive from having devoured gas and neighboring stars in those crowded regions.

DOUBLES AND VARIABLES

These basic stages that stars go through in life and death are fascinating. But there are other equally intriguing aspects of the lives of many stars. These stars may be variable stars (in fact, most if not all will be at some stage in life). They may be part of a double star or multiple star system (at least half of all stars are members of such a system). They may be members of a star cluster, moving group, or star association. We may see the nebula they are being born from, or the nebula they produce as they are dying. And we also see vast and distant systems of them that we call galaxies. Most of these things— double stars, variable stars, star clusters, nebulas, galaxies—we noted the existence of in the previous chapter and stated that there were a few examples of each visible to the naked eye. But the vast majority of them can only be seen through telescopes. So along with the explanation of what they really are and their place in the universe, we will here consider the appearances of these objects in the telescope and identify the terms you need to know to observe them skillfully.

Double Stars

Double stars may be physical doubles orbiting around each other, called *binary stars*. Or they may be at greatly different distances and only coincidentally appear near each other in the sky—*optical doubles*.

Stars that are too close together to have ever been separated visually in any telescope have had their "doubleness" detected by a spectroscopic examination and are called *spectroscopic doubles*. The brighter of the two stars in a double star system is usually also more massive and is called the primary, with the other star being the secondary or companion (if there are more than two stars, of course, all of the lesser ones could be called companions). Incidentally, the term *double star* is often used to include systems of three, four, or even more stars, though such a system may also be called a *multiple star*.

When observing a double star, the brightnesses of the components is important to know, especially because the dimmer a companion star is compared to the primary, the harder it will be to detect if it is quite close to the primary. Stars of about equal brightness are the easiest to separate and see. Also important is the amount of angular distance between the component stars. Here we rely on the smallest units of angular measure we learned in the previous chapter—arc-seconds (symbolized by "). There is a formula which predicts the *Dawes limit* as to the closest-together pair of stars you can separate with a telescope of a certain size. The key feature of the telescope in this respect (as in others) is the diameter of the primary mirror or lens in the telescope. The least separation resolvable with different lens or mirror diameters is as follows:

1-inch separates 4.56" (arc-seconds)
2-inch separates 2.28"
3-inch separates 1.52"
4-inch separates 1.14"
6-inch separates 0.76"
8-inch separates 0.57"
10-inch separates 0.46"
12.5-inch separates 0.36"
16-inch separates 0.29"

These figures have on occasion been exceeded by the sharpest-eyed and most skillful observers, but beginners will probably struggle to do as well. It is essential to realize that these are the resolving limits of telescopes when our atmosphere is steady enough to permit them. In other words, you will experience nights when the "seeing"—degree of steadiness of images, due to turbulence in our atmosphere—is poor enough to allow say only 1" or 2" resolution. On such nights a large telescope cannot split double stars or show details on planets or the Moon any better than a small telescope.

There is another statistic about double stars which is sometimes very important to know: the *position angle* or P.A. of a companion in relation to the primary star. We use the system in which 0° is north, 90° east, 180° south, and 270° west, back around to 360°, which is again 0°. I will rarely mention position angle in this book because most of the doubles discussed here are relatively easy to split. If, however, you are trying to see a companion star you haven't observed before that is very close to the primary, right at the limit of your telescope and "seeing" for the night, you will need to have the advantage of knowing the position angle.

The position angle can change over just a few decades or even a few years if a star has a fairly small orbit around its primary or, really, around a *barycenter*, a common center of gravity around which both stars revolve. So, too, can the angle of separation as seen from our vantage point. For a number of the double stars in the chapters ahead, I do give the separation figures at intervals for years or decades into the future.

Variable Stars

Quite a few variable stars can be enjoyed with the naked eye—especially if you are content to see them only near the bright end of their range. Binoculars and telescopes make observing even many of these fairly bright stars a lot easier and to their minimum brightness, and also make it possible to study the behavior of hundreds—even thousands—of dimmer ones. In this book, I am limiting coverage to the most interesting variable stars whose maximum brightness exceeds that which the naked eye can see.

The length of time between one peak (maximum) in brightness and the next is the period of the variable star. However, in some cases, there may be several maxima or minima in brightness. Also, even though some stars vary in brightness with tremendous regularity— they repeat their periods with clockwork precision each time—others are semiregular or even irregular.

One major classification of these stars is into *intrinsic variables* and *extrinsic variables*. An intrinsic variable changes brightness due to physical changes within the star. An extrinsic variable changes a star's brightness due to outside factors.

The common outside factor which changes the brightness of a star as we see it is an eclipse of one star by another. This happens when one member of a double-star system passes in front of the other as seen from our viewpoint. Such stars are called *eclipsing binaries*. The brightest and most famous example is the star Algol in the constellation Perseus. Like most eclipsing binaries, its period is just a few days long.

We can divide intrinsic variables into two further classes: *pulsating variables* and *eruptive variables*.

One kind of pulsating variable is the *long-period variable*, whose periods are typically months or even years long. The most famous of these, and brightest of those whose brightness varies by many magnitudes, is the star Mira in the constellation Cetus. Like most long-period variables, Mira is a red giant which changes not only brightness but size during the course of its period. These stars are struggling with instabilities as they go through transitional stages in their life. Another important kind of pulsating variable is the Cepheid, named after the most impressive star of the class Delta Cephei. The pulsations of Cepheids are extremely regular and most have periods measured in days. As we saw a few pages back, these stars also exhibit the *period-luminosity relation*, which enables them to be used as "standard candles" with which to measure distances in the universe.

Eruptive variables include *supernovas* and *novas* but also *flare stars* (stars with short-lived bursts of brightness), *dwarf novas*, Coronae Borealis–type stars, and other kinds whose brightenings are caused by violent or explosive events in the star. Supernovas we have already mentioned (a few pages back) but *novas* are not milder versions of the same phenomenon. They can result from several causes but most notably from disruptions of stars that have matter falling into them from nearby companions. In the chapters ahead, we will meet these stars and other dramatic eruptive variables, including stars which have been called *recurrent novas* (their great unexpected and sudden increases in brightness occurring multiple times over decades) and *reverse novas* (their brightness suffers dramatic drops at unexpected times before returning to their more normal brightness).

LARGER GROUPINGS OF STARS

There are larger groupings of stars than multiple star stystems. Among these are star clusters, moving groups, stellar associations, and galaxies. The last of these, as we will see, are very different from the first three.

Clusters, Moving Groups, Associations

There are two quite different kinds of star clusters. The first type are *open star clusters* (also sometimes called *galactic star clusters*). The second type are *globular star clusters*.

Open clusters are far more common. Something like eighteen thousand of them must exist in our Milky Way galaxy. The "open" in the name refers to the fact that there is considerably much open space between the stars in such a cluster. The stars are, in other words, far more loosely gathered than they are in a globular cluster. There are also far fewer stars in open clusters than in globular clusters, between about one or two dozen to hundreds of stars. On the other hand, open clusters may contain a wider variety of stars, including some of the most luminous giants, and therefore can be quite impressive. In addition, they are so common that we find several of them within a few hundred light-years of Earth—spectacular large- and bright-looking clusters like the Pleiades—and very large numbers of them within a few thousand light-years. By contrast, the two globular clusters closest to us may be only slightly less than 7,000 light-years away.

Globular clusters are far fewer in number than open clusters, with only a few hundred existing in our Milky Way. They are also not concentrated in the equatorial disk of the galaxy where our solar system is located, as open clusters are. Instead, globular clusters form a vast spherical halo in all directions outward from the center of the galaxy. Some globular clusters are within a few thousand light-years of the Milky Way's center but many are at great distances—three or four times, even six or more times farther from the center than our solar system is. Yet each one of these globular clusters is a spectacular roughly globe-shaped gathering of tens of thousands or even hundreds of thousands of stars. The most famous globular cluster is probably M13, the Great Globular Cluster in Hercules, one of a number of them bright enough to be glimpsed with the naked eye (though only as slightly fuzzy and enlarged points of light). In a medium-size amateur telescope, some of the best globulars are amazing balls of hazy radiance through which is shot the glints of numerous individual stars.

The kinds of stars in globular clusters are typically different than those in open clusters. The ones in open clusters are *Population I stars* while the ones in globular clusters are *Population II stars*. Actually, from what astronomers know now, it might make more functional sense if the titles were reversed. Population II stars are the oldest, "first-generation" stars. Population I stars can be second or third generation in the sense that they have formed from the material ejected by supernovas or perhaps even from material which has been used and reused in several supernovas. From its chemical makeup, the Sun is thought to be a third-generation star.

The current age of globular clusters can thus be 10 billion years or older in a universe like ours, which may itself be between 12 and 13 billion years old. The age of open clusters is typically measured in millions of years because as the years turn to billions the less strong attraction between these less tightly gathered stars gets overcome, the cluster members being tugged on by other passing stars, clusters, and interstellar gas and dust clouds in our galaxy's equatorial disk. Globular clusters are not only more concentrated and gravitationally tight to begin with but also are often located far above or beyond the equatorial plane, often in lone grandeur. So their stars tend to stay tightly bound to one another.

On the other hand, there are configurations of stars which are looser than even open star clusters. A *stellar association* is a confederation of young stars which may be diverging rapidly from the place they were born. An *O-B association* consists of hot, young stars of spectral types O and B (blue-white or white in color). The term *moving group* is applied to older stars that were probably once a star cluster but which have drifted farther and farther apart from their original fellow cluster members.

Nebulas

A nebula was once any light in the heavens which looked like a permanent hazy patch of radiance. As telescopes improved from the seventeenth to twentieth century, many of the celestial objects previously considered nebulas were shown to be collections of individual stars. They were star clusters in our galaxy or, in some cases, galaxies of stars far beyond our own galaxy. There remain, however, many cloudy patches of light that really are just that: clouds of gas and dust in interstellar space.

Nebulas are associated with both the births and deaths of stars. *Diffuse nebulas* are the clouds from which stars are born. Stars are born when the gravity of passing stars, other clouds, or "density waves" compresses regions of gas. The densest pockets then contract further by gravity until their central regions are dense enough to produce nuclear reactions. *Planetary nebulas* are clouds which many stars throw off as they are dying.

Diffuse nebulas may be *emission nebulas* or *reflection nebulas*. The emission nebulas produce light of their own after being stimulated by the radiation from very hot stars within or nearby. The reflection nebulas are not close enough to hot enough stars and therefore shine only by the light reflected from stars. Emission nebulas in long-exposure photographs or in rather large telescopes appear pinkish from the glow of hydrogen gas or, in the brightest parts of the brightest emission nebulas, green by the radiance from doubly ionized oxygen. Reflection nebulas glow blue—on long-exposure photographs or subtly through sizable telescopes. The most famous nebula which shines largely by emission is M42, the Great Nebula in Orion. The most famous which shines entirely by reflection is perhaps the Merope Nebula lit by the Pleiades star Merope.

There is actually one other class of nebula we should consider under the heading of diffuse nebulas: *dark nebulas*. These don't glow at all but are visible because they are seen in silhouette or profile against shining, diffuse nebulas or starry backgrounds. The most famous dark nebulas are the Coalsack near the Southern Cross and the Horsehead Nebula, which is marvelously prominent on photographs but elusive when sought visually in telescopes.

Completely different than the diffuse nebulas are the planetary nebulas. Their name is misleading. They really have nothing to do with the formation of planets or other aspects of planets. These nebulas were simply given this name because many of them appear blue-green, roundish, and small in telescopes and thus bear a superficial resemblance to the appearance of the planets Uranus and Neptune in such optical instruments. Planetary nebulas are in reality clouds of gas ejected from white dwarf stars in their long process of dying. Such a nebula emits its characteristic wavelengths of light in response to the hard radiation from the very hot surface of the white dwarf *central star* that ejected the nebula in the first place. Planetary nebulas last for much shorter periods of time than most deep-sky objects, possibly only tens of thousands or even just thousands of years. The brighter planetary nebulas are intriguing sights for amateur astronomers and in some cases their central star can be viewed in fairly small telescopes.

Galaxies

A *galaxy* is an immense collection of typically billions or hundreds of billions of stars. Our own galaxy we call the Milky Way. We see encircling the heavens on a dark country night a band of soft radiance that we also call the Milky Way—because that band is the combined light of distant stars that are located in the equatorial disk of our galaxy in which we ourselves are situated. Every individual star we see with the naked eye and almost all with telescopes is a member of the Milky Way galaxy. But even fairly small telescopes can show hundreds of patches of light which are galaxies other than our own.

Our galaxy has an equatorial disk because it is a *spiral galaxy*, shaped like a pinwheel or a Frisbee or a lens, or, as one astronomy writer wryly noted, like a fried egg with a central hub of billions of old (Population II) stars like the yolk of the egg. The spiral arms of a spiral galaxy may be either tightly or loosely coiled. Within those arms shine Population I stars, including many bright blue giants and red giants, great stellar associations, and regions of diffuse nebulas in which stars are being born.

Two other basic forms of galaxy are *elliptical galaxies* and *irregular galaxies*. Most elliptical galaxies are relatively small but some are the most gigantic galaxies of all. In some ways they seem to be like immense versions of globular clusters, with mostly Population II stars and comparatively little gas. Irregular galaxies are ragged and small but in some cases seem almost like detached sections of spiral arms, bright with hot, young giant stars and big bright nebulas. Large galaxies often have satellite galaxies that are irregular. Our own Milky Way galaxy has a number of satellite galaxies, the most prominent ones being irregular galaxies known as the Magellanic Clouds. These huge glowing patches are located at a high southerly declination and so are only properly visible at rather southerly locations on Earth.

Like stars, galaxies themselves can congregate. There can be smaller gatherings of them like the Local Group, which includes our Milky Way, the Great Galaxy in Andromeda, and a few dozen lesser members. Or there can be vast *galaxy clusters* like the one called the Virgo Galaxy Cluster, which consists of thousands of members.

Figure 2.2. An edge-on view and overhead view of the Milky Way galaxy, showing the location of our solar system. From the solar system's position within the galaxy's equatorial disk, we see that disk as a band of glow which encircles our heavens.

The Big Bang, the Expanding Universe, and Quasars

As most people have heard, the prevailing scientific theory about how our universe came into being is the big bang. In the early twentieth century, astronomers found that lines in the spectra of galaxies seemed to be displaced to the red end of the spectrum—or "redshifted." The only explanation for this which seems to fit the facts is that the universe—space itself—is expanding. This is what is causing all the galaxies in the universe (except those in groups whose gravity binds them strongly to one another) to be heading away from one another. The farther away a galaxy is, the faster the sum of our movement away from it and its movement away from us and the farther back in time we see it. As we begin the twenty-first century, evidence continues to support the idea that a little less than 13 billion years ago all present matter and energy of our universe were compressed into a tiny space which began its expansion in the event called the big bang.

As they image galaxies billions of light-years away with instruments like the Hubble Space Telescope, astronomers record earlier, more primitive forms of these objects. They also see objects called quasars much smaller than galaxies, emitting vast amounts of light and other energy. A few of these can be glimpsed in medium-size to large amateur telescopes. *Quasars* are now believed to be the stupendously luminous, superenergetic cores of galaxies that were active earlier in the history of the universe. The source for their tremendous energy output is believed to be massive black holes.

Cataloging Deep-sky Objects

Pulling our minds back from the farthest reaches of space and time, we return to practical questions about observing clusters, nebulas, galaxies, and even quasars through our backyard telescopes. The term to describe any of them is *deep-sky object*. A deep-sky object is any object beyond our solar system—though in practice, the term is often not applied to variable stars, double stars, or single stars. The first important thing to learn about these deep-sky objects is the designations astronomers have given to them and the catalogs astronomers have made to list them.

The most famous collection of deep-sky objects is the *Messier catalog* of more than one hundred Messier objects (or M-objects, for short) that was compiled by the eighteenth-century French astronomer Charles Messier. The Messier (pronounced MESS-ee-yay) objects were orignially intended to be a list of celestial sights which look like glowing fuzzy spots in telescopes and can be mistaken for comets—the objects Messier himself was most interested in discovering. Ultimately, Messier added to his catalog objects which were well known since ancient times and did not look like comets—for instance, the Pleiades star cluster, which even the naked eye shows plainly to be a group of stars. The entries in the Messier catalog receive designations like M1 (the Crab Nebula) and M45 (the Pleiades), up to M110 (though some traditionalists insist on limiting the number to the 103 that were published by Messier in his lifetime).

Charles Messier did discover comets but his greatest fame today is as the maker of this catalog of more than one hundred of the clusters, nebulas, galaxies, and other deep-sky objects that look brightest and best in amateur telescopes. By the late nineteenth century, however, astronomers needed a larger official catalog, so almost all the Messier objects are part of a listing of thousands of deep-sky objects compiled in the *New General Catalogue* (NGC). Many of the most beautiful and prominent clusters, nebulas, and galaxies are known by NGC designations like NGC 4755 (the Kappa Crucis Cluster or Jewel Box), NGC 7009 (the Saturn Nebula), and NGC 5128 (the galaxy Centaurus A). Further catalogs—like the two *Index Catalogs* (IC) published as supplements of the NGC—supply the designations for still other fascinating objects for amateur telescopes.

The chapters which follow are filled with deep-sky objects with colorful nicknames—the Wild Duck Cluster, Sombrero Galaxy, Helix Nebula. But the pages are also replete with combinations of letter-and-number designations (M11, M104, NGC 7293) which lovers of the deep sky come to know as affectionately as the celestial wonders themselves.

PART 2

The Months of the Stars

Chapter 3

December

[Constellations covered: Perseus, Taurus, Eridanus]

You can read this book straight through from beginning to end. Or, if you wish, you can plunge into it at any month—including the one that the world is currently experiencing. Still, despite our month-by-month format, the best way to organize this book into larger sections is by season. In this scheme of things, December is really the first month that the traditional constellations of winter are essentially all above the horizon at midevening. So we begin our tour of the year of stars with winter and with our chapter on December, even though it is the final month in the year of the civil calendar.

December is, for most of us at midnorthern latitudes, a cold and cloudy month. But when the sky does clear on December nights, such wonders shine in the heavens as would be worth an entire book of their own.

The brightest constellation of them all, Orion the Hunter, climbs the southeast sky on December evenings. Our map for the month also shows that the brightest of all stars, Sirius, is probably high enough to be seen sparkling like mad with glints of many colors, between the trees or buildings to your southeast. Other bright stars and patterns—Procyon in Orion's Little Dog, the famous Gemini the Twins, the much less famous but bright Auriga the Charioteer—are all well up in the east. All these constellations will be higher and better displayed in the midevening sky in January or February. So for December we will focus our attention on three constellations that lead the transition from late fall to early winter: brightly glittering Perseus the Champion or Hero; Taurus the Bull with his first-magnitude-star eye of orange and his handsome triangle of face and horns; and a far less famous constellation which is the longest north-to-south and holds special marvels: Eridanus, the River that runs right up to Orion's leading foot.

Within Perseus and Taurus we will meet some of the year's greatest stellar wonders. We'll encounter the brightest, most famous, and most dramatic of stars whose brightness varies in the course of a night, a star whose show is the eerie winking of a demon. We'll visit the telescope's most spectacular duo of side-by-side star clusters—twin piles of fabulous jewels. We will also learn about the two most prominent and beloved naked-eye star clusters in all the heavens—one of which is so compelling that it has appeared in humankind's greatest literature for thousands of years and has even been, in some cultures, the basis for their calendars.

PERSEUS

Andromeda and Cassiopeia are bright constellations we will study in detail in the November chapter of this book. They represent a famous daughter and mother, princess and queen, in Greek mythology. They also are neighbors in the sky to a male constellation no less bright and fascinating. The long axis of both Andromeda and Cassiopeia point to him and that is appropriate—for he is the center of the great myth to which they belong. His name is Perseus.

Andromeda is never far from her rescuer Perseus. (Check the December map in the back pocket of this book.) Her slightly curving line of bright 2nd-magnitude stars points to one of his two 2nd-magnitude stars—in fact, this star of Perseus is not much farther from Gamma Andromedae than Gamma is from Beta Andromedae or Beta from Alpha Andromedae. Alpha Persei is therefore almost a perfect continuation of the Andromeda line. And Beta Persei, the other 2nd-magnitude gem of Perseus, is not too far south of the line from Gamma Andromeda to Alpha Persei, and forms a right triangle with them. As we will see later in this discussion, Beta Persei is the famous variable star known as Algol.

The axis of the body of Perseus lies almost perpendicular to that of Andromeda in the heavens but more or less parallel to the long axis of Cassiopeia—and to the Milky Way band in which both Perseus and Cassiopeia lie. A line drawn through Gamma and Delta Cassiopeiae and extended southeastward points to the similarly oriented head and body of Perseus, with Alpha Persei in his chest or stomach. That line from Cassiopeia also passes through a spectacular deep-sky object which shines within the bounds of Perseus, a little above the hero's head. About halfway between the main patterns of Cassiopeia and Perseus glows this object, or duo of objects: the Double Cluster of Perseus.

The Double Cluster

Either of the two components of the Double Cluster would by itself be considered among the finest open clusters in the heavens. Together, they form one of the most admired deep-sky sights of both autumn and winter: I say two seasons because their high northern declination guarantees that they remain in sight from midnorthern latitudes for many months. In fact, at a declination of +57°, the Double Cluster is technically always above the horizon for observers north of 33° N latitude. In practice, of course, we want to see the Double Cluster when it is reasonably high in the evening hours, so fall and winter are the seasons to observe it.

The Double Cluster has been known since antiquity, being mentioned as far back as Hipparchus (who observed at Rhodes in 146–127 B.C.E.). The Double Cluster may be no more (or even a little less) prominent to the naked eye than the long smudge of M31, the Great Andromeda Galaxy, if you have an extremely dark sky. But in merely modestly dark skies it is more prominent to the naked eye than M31. Its position between the main patterns of Cassiopeia and Perseus helps increase the chances the eye will come across it. The Double Cluster looks like an elongated fuzzy glow in passable observing conditions, but on nights of both transparent air and good "seeing" even my not-very-sharp vision makes out a sort of fuzzy barbell or 8 shape, though not quite two entirely separate objects. It's puzzling why a Greek letter and another letter designation—"chi Persei" and "h Persei" were assigned to these clearly unstarlike objects as if they were stars. Also puzzling is why Messier didn't include the Double Cluster in his famous catalog. Could it be that he left it off his list because it was not discovered by him or his contemporaries and was already well known? Then why did Messier include even more famous clusters known since antiquity, like the Pleiades and Praesepe (Beehive Cluster)?

In small telescopes, the Double Cluster is impressive and in medium-size telescopes at low power the sight is magnificent. Twin piles of stellar jewels glitter side by side with lines of additional stars from them reaching out to each other. Perhaps statistics can convey on the page what otherwise only actual observation could supply (and of course surpass—this is a masterpiece of the heavens which must be seen, not merely read about!). The oft-quoted figures of magnitude 4.3 and 4.4 for the two individual clusters NGC 869 and NGC 884 are clearly too dim. Steve O'Meara's estimates of 3.5 and 3.6 are surely more accurate. The two clusters are centered about 1° apart and are each listed as being 30′ (30 arc-minutes) across but this latter figure is probably an underestimate, too. How many stars are there in NGC 869 and 884? There are about four to six dozen in each that are brighter than magnitude 11. But a professional study found that to dimmer limits there are about 200 stars in NGC 869 and about 150 in NGC 884. There are also plenty of foreground stars adding to the richness. Another study found that down to magnitude 12 there were, within 30′ arc-minute-wide circles centered on the two clusters, about four hundred stars (counting cluster members and noncluster members) seen in NGC 869 and about three hundred in NGC 884.

By the way, don't forget that you need a field of view well over a degree wide to get the main masses of both clusters together in one view. That means low power. Much higher power tends to make even clusters as relatively rich as these look less impressive.

It is the brighter stars in each of the clusters which stand out like featured gems in the twin hoards. I count a combined total of thirty-one stars brighter than 9.5 in the Double Cluster. According to Robert Burnham Jr., ten stars in the Double Cluster are brighter than magnitude 8.55. *Sky Catalogue 2000.0* lists the brightest star in NGC 869 as 6.55 and the brightest in NGC 884 as 8.05.

At first look, the similarity of the two clusters that make up the Double Cluster is striking. But further observation makes one also appreciate the intriguing differences. NGC 869 contains more stars and is richer, including being richer over a larger part of itself. It has no red stars, however. NGC 884 possesses some, including a prominent one near its center. Astronomers believe that one cluster is a modest distance farther than the other, and that they are of somewhat different ages—though both are certainly very young clusters, their hot luminous stars no more than about 3 to 6 million years old.

As spectacular as these clusters appear, they are even more impressive than they look. That is to say, the clusters appear as bright as they do even though they are both located more than 7,000 light-years away, in the Perseus Arm of the Milky Way, the next spiral arm outward from our own. What's more, astronomers estimate that their apparent brightness is dimmed by about 1.6 or 1.8 magnitudes by interstellar dust. When this factor is taken into account, it turns out that the ten brightest stars of the Double Cluster are comparable in luminosity to Orion's mighty beacons Betelgeuse and Rigel. Each cluster's total light may be about two hundred thousand times as great as our Sun's. Yet each is only about 70 light-years wide.

Interestingly, the Double Cluster is merely a part of a much bigger confederation of stars, the Perseus O-B1 Association. This O-B association is centered just west of the Double Cluster and covers an area about 6° wide. Scanning this region with binoculars and telescope does show it to be a fantastically rich one.

Other Star-Groupings and Nebulas in Perseus

Besides the Double Cluster, Perseus offers another naked-eye star cluster and several other kinds of stellar groupings bright enough to be seen with the naked eye. The grouping which is unquestionably an open cluster is M34. M34 shines at magnitude 5.2 and is an easy naked-eye sight in fairly dark skies. It lies about 5° northwest of Algol (Beta Persei), almost straight above the star as Perseus climbs the northeast sky. M34 is a quite sparse cluster but its individual stars are bright (for telescopic objects). It is definitely best seen at the very low magnifications of small telescopes or large binoculars.

Another stellar gathering in Perseus is sometimes classified as an open cluster, other times as the more loosely bound gathering called a moving group. This Alpha Persei Cluster or Alpha Persei Moving Group is a spectacular collection of bright stars extending around, but mostly for several degrees southeast from, Alpha Persei (Mirfak). In dark skies, the naked eye can show an impressive gathering but binoculars are really required for seeing the cluster in detail. Mirfak glitters at magnitude 1.8. Other prominent stars in the cluster include Psi, 29, 31, and 34 Persei. Sigma Persei is not a member but Delta and Epsilon Persei probably are outlying members. According to Kepple and Sanner's book *The Night Sky Observer's Guide*, 10 x 50 binoculars show forty stars visible within a 3°-wide core region. This core is more than 30 light-years across at the cluster's estimated distance of 540 light-years. Binoculars and very low power rich-field telescopes are the optimum instruments for a cluster as wide in the sky as this one.

A more widely scattered arrangement of stars than Alpha Persei's is the Zeta Persei Association. Zeta Persei is the final star of the leg of Perseus, which points right at the glorious Pleiades star cluster in Taurus. The Zeta Persei Association includes Zeta, 40, 42, Omicron, and Xi Persei. The first four of these stars in this O-B association can be fit into a wide telescopic field of view, but Xi is farther away. In all, there are more than twenty stars in the Zeta Persei Association and the remarkable thing is that their motions show a common origin point that they must have left little more than a million years ago. Thus we are seeing young stars in the process of rapidly escaping from their original mutual attraction for each other. The Zeta Persei Association is more than 1,000 light-years away and about 100 light-years across. At its northern edge, just north of Xi Persei, is an elusive but marvelous 3°-long strip of glowing gas known as the California Nebula. The shape really is somewhat like that of California. But you may have a hard time glimpsing it unless your sky is clear and dark and you use nebula filters and low-power optical aid. On very clear, dark nights, I have seen it dimly with a 10 x 83 finderscope, even without a nebula filter.

A much easier nebula to see in Perseus is M76, the Little Dumbbell Nebula. This planetary nebula's shape is, at first glance, roughly reminiscent of the much brighter and more famous M27, the Dumbbell Nebula in Vulpecula (see our September chapter). M76 is within a degree of magnitude-4 star Phi Persei, very near the border of Andromeda. Phi Persei and M76 are located just off of the line from Algol (Beta Persei) to Almak (Gamma Andromeda), about two-thirds of the way along the line. M76 is one of the most underrated of all Messier objects. Its magnitude used to be underestimated in books. It is at least a 10.1 magnitude object. That isn't very bright, of course, but M76 can be enjoyed in four-inch telescopes in a good dark sky. And an eight-inch telescope begins to show remarkable detail in this 67″ arc-seconds-wide ghostly exhalation of a dying star.

Algol the Demon Star

Don't spend all your time at Perseus with your eye staring into a tele-scope. The naked eye is what you need to appreciate the brightest of all variable stars whose changes in brightness are regular and short-period enough to enjoy frequently. Along with the slower, much longer-period variable Mira, this star is one of the two most famous variable stars in all the heavens. Its name is Beta Persei but people almost always call it Algol.

Algol usually shines at magnitude 2.1, making it the second-brightest star in Perseus. But every few days the star undergoes a rather precipitous dip in brightness to magnitude 3.4 and is outshined by at least five stars in the constellation. This minimum is reached every 2 days, 20 hours, 48 minutes, and 56 seconds, and occurs after a five-hour period of fading, which is followed by a five-hour period of brightening back to magnitude 2.1.

What is the cause of this striking behavior? Algol is the brightest of the class of stars called eclipsing binaries. The Algol system contains two stars, too close together to be separated with any telescope. When the dimmer star moves in front of the brighter, it eventually covers about 79 percent of the latter, bringing the point of light we see to minimum brightness. By the way, midway between these deep dim-mings there actually is a very slight dimming (too slight to notice visually) that occurs when part of the fainter star is hidden behind the brighter one.

The first recorded recognition of Algol's variability was made by Geminiano Montanari in 1667, and the regularity of the period was first determined by John Goodricke in 1782. It may seem strange that we have no record of Algol's shows from ancient times, considering how bright the star is and how drastic its dimmings are. Those dim-mings are not quite as easy to catch by chance as you might think, how-ever. You could go out on a number of nights and, depending on spells of bad weather, and Perseus being above the horizon only part of the night, keep missing the show for a long time. In order to guarantee you will catch Algol in action as soon and often as possible, check the list-ings for its minima (times it reaches minimum brightness) in *Sky & Telescope* magazine or the Web site SkyandTelescope.com.

Algol is an Arabic name which means "the ghoul." It was applied about a thousand years ago following Ptolemy's placement (nearly a

ALGOL
Light Curve

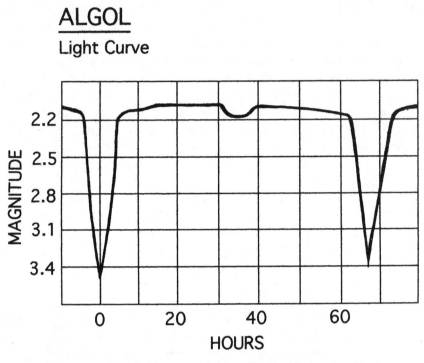

Figure 3.1. The light curve of the eclipsing binary star Algol.

thousand years earlier) of this star in one of the eyes of the dreadful monster Medusa. Ptolemy was merely following the ancient depiction of this part of the constellation as the severed head of Medusa being carried by hero Perseus. Did the eerie behavior of Algol influence ancient Greek skywatchers to make this star the eye of the monster? Maybe or maybe not. But even if not, the placement seems very appropriate to us today. Also appropriate is another title of Algol: the Demon Star.

The Myth of Perseus and Andromeda

Perseus was the son of a mortal woman and Zeus, the king of the gods. The young man found himself sworn by oath to attempt a perilous task—to bring back the head of the monster Medusa. Medusa was one of the three Gorgon sisters. Each Gorgon had wings, bronze talons, snakes for hair, and a face that was literally petrifying: one look at the face of a Gorgon would turn a person into stone. Perseus was given marvelous gifts by the gods but it would be up to him to use them with intelligence and courage. He flew to the island of the Gorgons on the winged sandals he'd borrowed from Hermes (Mercury), messenger of the gods. He was wearing the cloak of invisibility given to him by Athena, goddess of wisdom. But even if the Gorgons didn't see him, how could he dare look toward them, even just long enough to cut off and grab Medusa's head? How could he slay Medusa when she might at any moment turn her dreadful visage in his direction and convert him into a statue, frozen with a last look of horror on his face? Perseus flew down on the sleeping Gorgons looking only at the images of them in his shield. The images apparently did not have the petrifying power because they were distorted. He swooped in, chopped off Medusa's head, put it into a bag, and lifted off. At first, the flying monster sisters were hot on his heels. But they could only listen and grasp for him, not see him, and he soon escaped. The winged horse Pegasus leaped from the spilled blood of Medusa, a beautiful marvel springing from a horrible one.

Meanwhile, far away, King Cepheus and Queen Cassiopeia were parents of the innocent princess Andromeda, but when Cassiopeia boasted that she (Cassiopeia herself) was as beautiful as the Nereids, these sea nymphs asked Poseidon, god of the sea, to send a terrible monster to ravage the coasts of the land. He did so and the monster's attacks were devastating. An oracle that the king and queen consulted said that the only way to end the attacks by the sea monster was to let poor Andromeda be devoured by the beast. And so Andromeda, chained by the ocean, watched in terror as the sea monster appeared and headed toward her.

Fortunately, Perseus was flying past and saw the horrible scene below. Just in the nick of time, Perseus swooped down. He dove at the sea monster, using himself as bait and taunting it far enough away from Andromeda and the shore so that he could use his ultimate weapon. Averting his vision, he pulled out the head of Medusa and held it toward the sea monster. The beast instantly turned into stone—an island offshore which people can still see to this very day. Putting the terrible head back into the bag, Perseus flew to shore and freed the chained Andromeda. The two fell in love, were married, had children, and lived happily for the rest of their lives. After death, their forms were even translated up into the heavens into the form of splendid constellations. In addition to Perseus and Andromeda, the parents of Andromeda (and also Pegasus) made it to the stars. So, too, did the sea monster, but at least he is well separated from Andromeda and has become known as a whale—Cetus. And Perseus is still nearby to come to Andromeda's rescue.

TAURUS

The main pattern of Perseus is something like a big cursive letter K when it rises or, when it is higher, like a hand with Algol marking the thumb and other stars marking the fingers hanging south. If so, what does the hand seem to be reaching to pick up? A gathering of stellar gems so prominent by virtue of its brightness and compactness to the naked eye that it is truly the most unique starry sight in all the heavens. That pocketful of jewels is the Pleiades star cluster.

The Pleiades were seven sisters in Greek mythology, the daughters of Atlas and Pleione. The star grouping is often called the Seven Sisters, even though only six of the stars are fairly easy to distinguish with the naked eye in moderately good sky conditions, and more than seven are visible to sharp unaided eyes under superb conditions. The Pleiades almost deserve to be considered a tiny constellation in their own right. Since they look to the naked eye somewhat like a miniature dipper with short handle, many people who see them think they are the Little Dipper (which is actually much larger and far less conspicuous than the Pleiades). But the Pleiades are not a constellation. They lie at the shoulder of a true constellation, one of the most famous of the zodiac—Taurus the Bull.

We will return to the Pleiades and all their beauties and lore in a few minutes. First, however, let's study the pattern and major stars of Taurus itself.

The Stars of Taurus

Taurus lies midway between Perseus and Orion. But the best way for a novice to locate it is to begin with Orion's amazingly prominent Belt of three similarly bright stars in a row. If you extend the line of the Belt far to the the upper right (northwest) on December evenings, you'll find it takes you near to a 1st-magnitude orange star that is part of a remarkable pattern of moderately bright stars.

The 1st-magnitude star is Aldebaran. It forms with those lesser stars a prominent letter V or arrowhead pattern. This V outlines the handsome face of Taurus the Bull. Aldebaran, the top (northern) end of one of the arms of the V, marks the eye of Taurus. Aldebaran is an appropriate bull's-eye, with its orange color at least suggesting the bloodshot red in the glare of an angry bull. In mythology, Taurus is most often associated with the bull form that Zeus, Greek king of the gods, assumed to carry off the maiden Europa. (By the way, Europa has had her name applied to one of the major moons of Jupiter, a moon which is believed to have beneath its icy surface an ocean of water that just possibly may contain life.) The placement of the constellation Taurus facing constellation Orion has suggested a story of its own—a confrontation between the two, Bull and Hunter, here in the heavens. Aldebaran shines at magnitude 0.87, usually somewhat fainter than Orion's redder star Betelgeuse.

But where are the horns of the Bull? These are only marked with stars at their tips and those particular stars are more than a fist-width at arm's length to the upper left (northeast) of the Bull's face. If you extend the arms of the V of the Bull's face you will find that they approximately point you to the horn stars. Another way to locate the horn stars—Zeta Tauri and Beta Tauri—is to note that they lie about halfway between the brilliant yellow star Capella in Auriga and Betelgeuse in Orion.

Beta Tauri is the more northerly and much brighter of the two horn stars. It shines at magnitude 1.7 and is known as El Nath or Nath, from the Arabic for "the butting." Beta Tauri officially belongs to Taurus but it is unofficially used to complete the pentagon shape of the main pattern of Auriga the Charioteer. Thus the patterns of Taurus and Auriga are linked together by this star.

Zeta Tauri is the more southerly and much dimmer of the two horn stars. Its magnitude is only 3.0. But it does have the distinction of being closer to the ecliptic, the midline of the zodiac, and thus more often has a planet or the Moon pass near it. Another distinction of Zeta is its proximity to one of the most fascinating of all nebulas—the Crab Nebula, which we will return to later in this chapter.

Aldebaran and the Hyades

Let's return first to Aldebaran and the V of stars to which it belongs. The other stars of this V, along with a number of additional stars around and near them, are actually members of the Hyades star cluster.

Aldebaran lies only 65 light-years from Earth, more than twice as close as the Hyades cluster, which is located 150 light-years from us. There is only one star cluster (the Ursa Major Cluster which includes some of the Big Dipper stars) which is closer to us than the Hyades, but it is difficult to glimpse all the widely scattered members of that group at once and to know which stars among them are nonmembers. In contrast, the Hyades form a single discrete sight. The only non-member which you might easily mistake as being one of them is Alde-baran, due to its place in forming the V or arrowhead. All but one of the six brightest Hyades stars glows at between magnitude 3.4 and 3.8. Many dimmer stars populate this cluster, which appears so large—more than 5° across—that it needs a binoculars' wide field of view to contain it all. There are some beautiful pairs of stars in the Hyades. Theta^1 and Theta^2 Tauri are one of the interesting naked-eye tests of vision, a magnitude 3.4 and 3.8 star just over 5 1/2' apart. Notice in binoculars or a small telescope the different colors of the Hyad stars. The cluster is old enough so that the blue-white stars of spectral type A are joined by a number of red giants. Three of the five brightest cluster members are K-type and another is a G-type giant. Fifteeen members of the Hyades cluster are brighter than 5.0! It is an even brighter cluster than the Pleiades but is spread out a lot more than the Pleiades (being much closer) so that the naked-eye view is less eye-catching.

Aldebaran is an orange giant which is probably about twenty times as wide as the Sun. The ecliptic runs not far north of it and the Hyades, so the Moon sometimes passes through the cluster and right over some of its members—and over bright Aldebaran. The name Aldebaran is from the Arabic for "the follower"—not the follower of the Hyades but rather the Pleiades, in the journey across the sky each night.

The Pleiades

No compact pattern of stars in the sky except maybe Orion's Belt is more conspicuous than the Pleiades. But the Pleiades is more con-densed than the Belt (the main stars are huddled within about 1 1/2° of each other), and more numerous and mysterious-looking. Poor vision or poor sky conditions can make the cluster appear like almost a single odd mass of light. Twinkling of stars is caused by cells of tur-

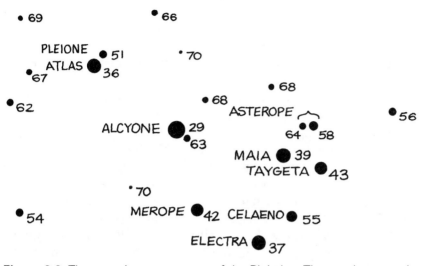

Figure 3.2. The most important stars of the Pleiades. The numbers are the magnitudes of each star with the decimal point (which might be mistaken for a dim star) removed—thus "69" means magnitude 6.9.

bulence in our atmosphere, but these cells are generally too small to affect the entire Pleiades at once. So we have the unique and thrilling experience of seeing waves of twinkling pass through the cluster.

The Pleiades is such a lovely and gentle sight that most cultures have imagined these stars to be innocent maidens or doves or baby chicks gathered around a mother hen (the brightest star of the Pleiades). The cluster is one of the few starry sights mentioned in the Bible and in Homer's writing. Moving almost three thousand years ahead to the present, we can see a stylized image of the Pleiades in the emblem of Subaru automoblies (Subaru is the Japanese name for the Pleiades). The great nineteenth-century poet Alfred Lord Tennyson wrote in famous lines of "Locksley Hall": "Many a night I saw the Pleiads, rising thro' the mellow shade,/Glitter like a swarm of fireflies tangled in a silver braid." This may be a reference to either the intermingling of star beams when waves of twinkling pass over the cluster or to naked-eye glimpses of traces of nebulosity around the Pleiades. These patches of glow can be glimpsed on good nights by a telescopic observer, especially in the region extending south of the Pleiad star Merope. The nebulosity is spectacular on long-exposure photographs of the cluster. It was once thought that these patches and strands of nebulosity around some of the Pleiades were remnants of their formation, but we now know the gas is unrelated to their genesis, merely being lit up by the hot stars of the cluster as they pass through it.

The Pleiades cluster is located 375 light-years away and its stars are all hot, young, rapidly spinning stars of spectral type B. The keenest naked-eye observers have seen up to eighteen stars in the cluster but binoculars show dozens and there may be a few hundred members visible in telescopes. The brightest Pleiads form a tiny dipper shape, though perhaps it would be more accurate to compare this short-handled structure to a teacup. The handle is made up of a pair of stars actually named for the parents of the Seven Sisters—Atlas and Pleione. The two are of magnitude 3.6 and 5.1 and are very close together, making them a hard split for the naked eye especially on nights of poor "seeing." What's more, Pleione is at least slightly variable. This has led to the theory that in the past Pleione was brighter and easier to see, so that people really did behold "Seven Sisters" more easily than in modern times when casual observers usually notice six. But the legend of a "Lost Pleiad" dates far back and is told by a number of cultures that are greatly separated throughout the world. In the nineteenth century, Alfred Austin expressed this legend succinctly: "The Sister Stars that once were seven/Mourn for their missing mate in Heaven."

The brightest star of the Pleiades is Alcyone, marking the junction of "handle" and "bowl" with its magnitude 2.8 light. It was once theorized to be the central sun of the universe! The other bright named stars of the Pleiades bowl, working counterclockwise from Alcyone are: Merope, Electra, Celaeno, Taygeta, and Maia, with Asterope— actually a wide double star—floating just north of Taygeta and Maia. The nine brightest Pleaides are all brighter than magnitude 5.6.

The Pleiades are the subject of countless legends around the world. One Native American myth tells of maidens who were chased up a mountain by a giant evil bear. The scratching claws of the bear turned the mountain into the strange landform with striations on its sides that is now called Devils Tower in Wyoming. The maidens were saved by being lifted up to the heavens to become the Pleiades. I have long wondered if Steven Spielberg knew of this tale when he decided to use Devils Tower as the launch place for a voyage to the stars in his classic movie *Close Encounters of the Third Kind*.

The Pleiades have also been the basis for calendars in some cultures. The Druids of Britain apparently celebrated their New Year with the holiday Samhain, which became the basis for the holiday we now know as Halloween. And that New Year began on the date when the Pleiades were seen to be rising at nightfall.

The Crab Nebula

We must visit one more object before leaving the confines of Taurus the Bull. A few pages back I noted that there was a remarkable deep-sky object near the fainter hornstar, Zeta Tauri. That object is the Crab Nebula, only about 1° northwest of the star.

The Crab Nebula is M1, the very first entry on Charles Messier's list of deep-sky objects. The Crab Nebula is the brightest of the supernova remnants (SNRs). This glowing—and still expanding—cloud of gas is the product of a mighty star explosion that became visible from Earth in July 1054. The Crab supernova was brighter than Venus for awhile. It was observed in China in broad daylight for twenty-three days. In the twentieth century, M1 became perhaps the most studied of all objects beyond our solar system. Astronomers identified the star that went supernova as a pulsar spinning thirty-three times per second.

Some beginners who have seen impressive photographs of M1 expect too much from their first views of it in the telescope. The Crab does appear as a rather featureless dim blur in small telescopes, even if sky conditions are not too bad. But with a little more aperture it becomes a far more interesting object. It glows at a magnitude of about 8.4, and measures 6' by 4'. With an 8-inch telescope in very good skies, a few of the scalloped edges of M1 become visible and can start being imagined as claws, showing why the nineteenth-century observer Lord Rosse compared the nebula to a crab. In larger telescopes and on photographs, the eerie filaments of the nebula can be readily glimpsed. The distance to M1 is thought to be about 6,500 light-years.

ERIDANUS

Many people have heard of Perseus and most have heard of Taurus, but how many people who are not already devoted astronomers have heard of Eridanus the River? It certainly does not have the visual spectacles to rival those of Perseus and Taurus. But it does contain a few very fascinating sights.

Proof that Eridanus held more interest in yesteryear, when light pollution was not a problem, is the number of proper names for stars in this mostly dim constellation. Cursa, Keid, Beid, Zaurac, Rana, Azha, Angetenar, Acamar, and Achernar: these are proper names of some of the Eridanus stars, starting from the river's northern end near the brilliant star Rigel in Orion. The other end of the river—the southern end—was once at the star Acamar (Theta Eridani), which lies at –40° declination. But it was extended another 17° farther south to include a star whose name, Achernar, is a variant of Acamar—both are from an Arabic term meaning "end of the river." Acamar shines at magnitude 2.9. But if you live far enough south you can see Achernar burn at magnitude 0.45, making it usually the 9th brightest star in all the heavens. This star, Alpha Eridani, lies 144 light-years from Earth.

One thing Acamar has in its favor is that it is a double star. If you live north of about 40° N. latitude you may struggle with bad "seeing" for any object so low in your sky, but for observers far enough south, Acamar is a pair of white stars, magnitudes 3.4 and 4.5, 8" (8 arc-seconds) apart. Dimmer than Acamar but much farther north in Eridanus (in fact, just south of the celestial equator) is the double 32 Eridani. This is a yellow star of magnitude 4.8 that is 7" from a blue-green star that is magnitude 6.1.

Eridanus also offers a dim triple-star system which includes the easiest of all white dwarf stars to observe. $Omicron^2$ Eridani (also known as 40 Eridani) features a magnitude 4.4 orange star with two companions: a magnitude 11.2 red dwarf 8" away and a magnitude 9.5 white dwarf 83" from the primary. The white dwarf companion of Sirius is brighter than the white dwarf $Omicron^2$ Eridani B, but the former is overwhelmed by its nearby brilliant primary. So this star in Eridanus is the easiest of its class to see. The $Omicron^2$ Eridani system is 16.5 light-years away from us, almost twice as far as the Sirius system.

There is a naked-eye star in Eridanus which is much closer than $Omicron^2$. This is Epsilon Eridani, which shines at magnitude 3.7 and is only 10.5 light-years away. It is a K2 star that is considerably smaller and less luminous than our Sun. It resembles a star not far across the border into Cetus, Tau Ceti. Tau is a G8 star of magnitude 3.5 that is 11.9 light-years from us. These two were the first stars listened to for possible extraterrestrial radio transmissions, more than forty years ago by Frank Drake, as part of Project Ozma, arguably the first important act of SETI (the Search for Extraterrestrial Intelligence). Today, there is fresh evidence that Epsilon Eridani may have life-sustaining planets.

DECEMBER'S OBSERVING EXPERIENCE AND GEMINID METEORS

The stars of December are practically matchless. But part of the observing experience just before midmonth in December is watching the Geminid meteor shower.

The Perseid meteor shower of August is the most famous and most frequently watched of meteor showers. In many years, though, the strongest and finest meteor shower is the Geminids.

The Geminids typically reach their peak on December 13 or 14 and in a good year (moonless sky, the half-day-long meteor maximum timed well for your longitude, etc.) the country observer with clear skies may see sixty or more Geminids per hour at peak. The meteors appear to diverge from a radiant point near the Gemini star Castor. They become numerous much earlier in the night than the meteors of most showers—even by 9 or 10 P.M.—because the radiant is already fairly high in the east by midevening. The radiant is highest—virtually overhead for midnorthern latitudes—shortly after 2 A.M., so numbers tend to be best around midnight and the next several hours.

The Geminids hold the distinction of being the only major annual shower believed to be derived not from a comet, but from an asteroid. (The asteroid is Phaëthon, which passes well within the orbit of Mercury at its near point to the Sun.) There is evidence that the numbers of Geminids may dwindle during the twenty-first century. This is yet another reason why we should not put off going out to see this marvelous shower, even if doing so means braving the cold.

December is the cloudiest month of the year in many lands at midnorthern latitudes. This is a time of year when the Pacific Northwest especially gets blanketed with overcast for much of the month. Even here in New Jersey I've seen several years when literally only two or three nights in the entire month were mostly clear.

The rarity of clear nights makes it all the more important that we go out to experience them, despite the cold. In the January chapter, I explain how to stay warm on winter nights while stargazing. I also explain why January tends to be the coldest month of the year in the Northern Hemisphere even though December is the month of that hemisphere's winter solstice. The winter solstice is the time on about December 21 when the Sun is farthest south in the heavens and takes its lowest arc across the sky, making these days the shortest of the year and these nights the longest.

December nights are not always severely cold (in New Jersey, I watched the Geminids in 0°F weather one year but then in 60°F the next!). But there is a temptation to spend the nights inside around a fire. That is certainly an appealing thing to do during December's holidays. We should remember, however, that we can also hold a celebration outside on December nights—a celebration of the heartwarming fires and soul-stirring flames of the heavens that are so bright at this time of year.

During the first half of evenings in December, one of the great monents for the naked-eye skywatcher is when he or she first glimpses Sirius as it rises. As I wrote at the start of this chapter, when Sirius is low in the sky, it is usually a veritable fountain of not only twinkling but also color changes. Yet there is a moment which comes later in the night that is also memorable: the moment that Sirius, most brilliant of all night's stars, stands at its highest in the due south. By a marvelous coincidence this happens exactly at midnight on New Year's Eve. One year when I was a teenager my brother and I walked along a fire trail through dark woods as the final evening of the year drew to a close. How did we spend the minute that the new year came in? We spent it looking from a clearing in those woods at the supreme star Sirius as it beamed down to us from the throne of its greatest height in the heavens.

Chapter 4

January

[Constellations covered: Orion, Auriga, Lepus, Columba]

Our calendar year begins with the brightest month of the stars. January is when the brightest constellation, Orion the Hunter, is at his highest in the south in the middle of the evening. It is also when the brightest group of constellations, which includes Orion, is at about its collective highest at that convenient time.

The thing which is not convenient about January evenings is, of course, their chill. On average, January is the coldest month of the year in most parts of the Northern Hemisphere. Why not December, when the winter solstice occurs, and days are shortest? Because Earth's climate systems (regulated in part by the huge oceans) lag behind, taking a month or so to react fully to receiving the minimum of sunlight.

A more surprising fact to most people is that sometime between January 1 and 4 each year is when Earth reaches perihelion—that is, it's closest to the Sun in space. How could we have our coldest weather around the time that Earth is closest to the Sun? First of all, we should remember that it is only Earth's Northern Hemisphere that has its coldest weather in January. In January, the Southern Hemisphere is experiencing summer. But northern lands get cold in January because the Northern Hemisphere is tilted away from the Sun, so that the Sun takes a low arc across the sky and is above the horizon for relatively few hours. The tilt of the Earth has a more important impact on temperature than the variations in our planet's distance from the Sun, because the latter are rather small. Earth at perihelion in January is still at more than 98 percent of its average distance from the Sun.

So January nights are cold in northern lands. How can a person dress warmly enough to stay comfortable while standing mostly still under the stars for a long time on a very cold night? On a different note, is there truth to the popular notion that the stars look brighter in winter because winter nights are more clear? The answers to these and other questions will be explored in this chapter. We will also learn how to see what is sometimes the strongest meteor shower of the year, why legend says that Orion the Hunter sets before Scorpius the Scorpion rises, and how we can look out the back window of our solar system.

What's most important for this journey in January, however, are the wonders of its prime constellations, especially Orion. Orion is not only the most dramatic constellation of all for naked-eye observers. It also contains what most observers think is the most spectacular of all deep-sky objects in a telescope—the Great Orion Nebula. When you read the following pages or, better yet, step outside to live them, you may find that there is a shiver running down your spine that is not from cold. The stellar splendors of January bring a chill and thrill of beauty.

ORION

When Orion reaches the halfway point in his journey across the heavens and is in his highest position, the human figure that his stars represent is standing upright. No figure in astronomy is more gallant than Orion above the winter landscape. Winter is not a kindly season but has an austere beauty. It seems appropriate that at such a time of year there is a warrior constellation decked out in bright stars that mostly shine with a biting hint of icy blue.

Why is Orion so spectacular? Not just because it is the brightest constellation. Orion also holds a marvelously central position in several ways.

First of all, the tall form of the Hunter stands at the center of a great semicircle or great arch of the other bright constellations of winter. Second, Orion's Belt lies virtually right on the celestial equator. The westernmost of the Belt stars is less than one-third of a degree south of the celestial equator in our era of history. This means that at least part of Orion is visible from every place on our planet.

More immediately obvious than all the forms of centrality which Orion has is the symmetry of Orion's pattern. What could be a more striking arrangement of symmetry than the three stars (southeast to northwest) called Alnitak, Alnilam, and Mintaka—the Belt of Orion? The Belt is the sole instance in all the heavens of three bright stars—similarly bright and similarly spaced—forming a short row (just 3° long). No other compact asterism is so eyecatching, unless it be the tiny dipper and bunching of stars which makes up the Pleiades star cluster. But the Pleiades stars are far dimmer than those of the Belt.

Yet the Belt is just the center of the symmetry of the larger pattern of Orion. The two stars about 10° above the Belt (Betelgeuse and Bellatrix) are considerably farther apart than the 3° length of the Belt and so resemble broad shoulders. So are the two stars about 10° below the Belt (Saiph and Rigel), which therefore are imagined to mark the mighty Hunter's knees. Most cultures have seen this pattern of seven stars—the three-star Belt, the two-star shoulders, the two-star knees—as irresistibly human. And because it is so bright, this human form is imagined to be a glorious hunter or warrior.

Orion is even equipped with a sparkling sword. The Sword of Orion is the north-south line of three modestly bright stars a few degrees south of the Belt. The Great Orion Nebula seems like a gleam of the Sword, almost hidden to the naked eye by the brightness of the middle star of the Sword. But a pair of binoculars or a telescope opens up the view of the Great Orion Nebula into first a puff of hazy light and then into a fantastical fan of intricate green radiance.

Orion's Bright Stars

Of course, even having marvelous symmetry, a form reminscent of a human figure, and being centrally positioned in several ways would not be enough to make Orion the most spectacular of the constellations if its stars weren't brilliant.

They are. Five of the seven stars of Orion's main pattern are second magnitude. Six of the seven stars of the Big Dipper are second magnitude. But the seventh is much dimmer. The sixth and seventh stars of Orion are much brighter than second magnitude. As a matter of fact, they rank as numbers 7 and 10 on the list of the brightest stars visible from Earth, and as numbers 5 and 7 on the list of brightest stars properly visible from the midnorthern latitudes where so much of our planet's population lives. The only constellations other than Orion to possess two stars of first magnitude or better brightness are the far-south constellations Centaurus and Crux. Centaurus as a whole, however, is sprawly and far from symmetrical. And the tiny Crux—the Southern Cross—often disappoints people because its fabled symmetry is missing any bright star in the center of its cross pattern.

Orion's two stars which shine much more brilliantly than the rest of the constellation are Rigel and Betelgeuse. They are marvelous complements of each other. Betelgeuse marks the east shoulder of Orion; Rigel marks the west knee. Betelgeuse shines a distinctive orange-yellow; Rigel shines slightly bluish white. Now it is true that star colors are subtle to the naked eye. And there are differences in not just color vocabulary but in actual physiology between different observers so that they may not agree on the exact tints of particular stars. But most people will perceive the difference in color between Rigel and Betelgeuse with the unaided eye, especially if they glance quickly back and forth between the two stars, which is easy to do.

Rigel and Betelgeuse are not just examples of a blue-white and an orange-yellow star. They are the brightest, classic examples of two of the most important types of star, types which we discussed in chapter 2: Rigel is a blue giant and Betelgeuse is a red giant.

Betelgeuse

Although Rigel is brighter than Betelgeuse in both apparent brightness and true brightness, Betelgeuse is an even more distinctive star. In fact, Betelgeuse might be the standout star in all the heavens if we consider its physical nature.

Betelgeuse is: one of only two 1st-magnitude stars that appear reddish (really more like the color of a pumpkin than the color of a tomato); one of only two such stars which is a red giant; and one of only two such stars whose brightness varies noticeably. The other example is the summer star Antares. Antares had long been considered slightly inferior to Betelgeuse in luminosity and size, but if it is about 600 light-years away, it may actually be a little more luminous than Betelgeuse. There is no doubt, however, that Betelgeuse has a greater apparent brightness—and larger and more frequent variations in brightness—than Antares. Even if Antares is more luminous, Betelgeuse seems to be a slightly cooler star than Antares and probably emits so much infrared radiation that it exceeds Antares in total amount of energy emitted. If our eyes were sensitive to all wavelengths of electomagnetic radiation, Betelgeuse would probably be the brightest star in all the heavens.

Betelgeuse may also be the *largest* sun within a few thousand light-years of Earth. At its distance of only 520 light-years, Betelgeuse certainly has the greatest apparent diameter of any star as seen from Earth. The stars are so remote that even in the case of Betelgeuse this works out to only about 0.05″ (arc-seconds), too tiny to be glimpsed visually with even the world's largest telescopes through Earth's ceaselessly unsteady atmosphere. But astronomers have used special techniques to cancel out the jitter of images recorded with ground-based telescopes. And above Earth's atmosphere, orbiting observatories can obtain steady images. Such an image of Betelgeuse produced by the Hubble Space Telescope revealed a brighter region in the image which is thought to be an actual surface bright spot on the star (other images of giant stars have revealed dimmer, cooler areas which seem to be the equivalent of sunspots—starspots—on these stars).

If we are correct about the distance of Betelgeuse, the aforementioned Hubble image indicates that the star may be well over one thousand times wider than the Sun. If transported to our solar system then, Betelgeuse would fill all the distance between the Sun and Saturn. Or, if we could magically replace our Sun with Betelgeuse, we would find that it engulfed our solar system out to the orbit of Jupiter.

Of course, in chapter 2 we discussed how even our own Sun when it becomes a red giant would, in a few billion years, swell out to the orbit of Venus and perhaps the orbit of Earth. But Betelgeuse probably has something like twenty times the mass of our Sun. And there is a fact about its size and power which is perhaps more amazing than any of the above: Betelgeuse has a companion star which travels at least part of its orbit *inside Betelgeuse.*

A star traveling within another star, or passing in and out of another star? It hardly seems believable. But we should remember how tenuous the outer layers of a red giant are and that this companion is surely a hotter, denser star. The drag on the companion star would be far less than we might think. Nevertheless, the companion will eventually have its orbit decay and at last be drawn deeper into Betelgeuse to merge with the greater star. This ultimate swallowing will occur soon by astronomical standards—perhaps as soon as mere

Figure 4.1. Position of Betelgeuse in Orion and Hubble Space Telescope image of star's actual globe.

thousands of years from now. Unless, of course, Betelgeuse goes supernova sooner than that.

Is Betelgeuse About to Blow?

We sometimes hear the claim that Betelgeuse is the star in our sky, or at least naked-eye star in our sky, most likely to go supernova tonight. Actually, if we saw Betelgeuse brightening spectacularly one night in, say, 2012, then the actual supernova would have erupted around the time Columbus first set sail for the Indies and "discovered" America. Maybe as we all watched it these past five centuries Betelgeuse no longer even existed, goes one story (actually there would always be a remnant of the star, though presumably it would be a black hole).

But even stars as massive and profligate of their fuel as Betelgeuse have a red giant phase probably measured in millions of years. So we are very unlikely to see Betelgeuse or any other star currently visible with the naked eye go supernova in the next few thousand years, let alone in our lifetime. Still, you can never tell. If Betelgeuse does go supernova when it is roughly as close as it now is to us, it would become much brighter than a Full Moon in our sky for a few months. A recent study argues that radiation from a supernova at that distance (from a star a few hundred light-years from Earth) would not cause a major extinction event on Earth like those produced by asteroid and comet impacts. This is a very speculative topic, however. Most of us would not want to be around to see the prediction put directly to the test.

Like many red giants, Betelgeuse does at present show dramatic signs of variability in brightness. Sir John Herschel seems to have been the first to notice the star's variability, back in 1836. In December 1852 Herschel apparently felt that Betelgeuse had brightened to negative magnitude, outshining stars like Capella and Rigel. In the period between 1937 and 1975 Joseph Ashbrook judged that Betelgeuse became as bright as –0.1 and as dim as +1.1. How bright does it look to you tonight? Remember that when comparing Betelgeuse to other winter stars like Capella, Procyon, or Aldebaran you have to factor in the altitude of the stars due to the dimming effect of Earth's atmosphere, "atmospheric extinction." Looking lower in the sky you are staring through a greater thickness of our atmosphere. Therefore, even on a good clear night with minimal haze, a star 10° high is dimmed by about 1.0 magnitude, a star 15° high by 0.7, a star 19° high by 0.5, a star 26° high by 0.3, a star 32° high by 0.2. And even a star 43° high—almost halfway up the sky—is dimmed by 0.1 magnitude.

The brightness variations of Betelgeuse are not regular but do seem to show a main period of about six years with superimposed secondary periods of about 150 to 300 days. The changes in brightness are usually quite gradual but Ashbrook noted a rise and then a fall of 0.4 magnitude in just two weeks in February 1957. This is yet another reason that Betelgeuse always bears watching.

Betelgeuse and many other red giants undergo variations not just in brightness but also in size. The star is like an ever so slowly pulsing vast heart, its diameter altering by perhaps as much as 60 percent, sometimes over a matter of months or less.

No discussion of Betelgeuse would be complete without a few words about its name. The pronounciation "beetle juice" has become known to much of the public and has occasioned more than a few laughs. In reality, most astronomers pronounce the proper name of Alpha Orionis "BET-el-joos." Of course, astronomy writers sometimes argue that because the name is a terribly muddled corruption of a phrase in medieval Arabic, there is no truly "correct" pronounciation. But dictionaries themselves are based upon the standards of prevalent usage among the frequent contemporary users of a word or name. (If you are interested in pursuing the origins of the name Betelgeuse and how the name would have been pronounced by Arab astronomers, you can consult the most authoritative work in English.)[1]

Rigel and the Orion Association

From Orion's most distinctive star, Betelgeuse, we turn back to its brightest, Rigel. Rigel is the classic example of a blue giant. It is thought to have a greater true brightness than any of the other 1st-magnitude stars except Deneb. Rigel was once thought to be one of perhaps the half-dozen most luminous stars in our section (say a 10,000 light-year-wide region) of the galaxy. However, more recent estimates of Rigel's distance judge it to lie about 770 light-years from us, closer than previously thought. As a result, Rigel is believed to have an absolute magnitude of about –6.6, similar to that of some of the other brightest blue giants in the Orion region. For instance, Rigel seems to have about the same luminosity as Alnilam (Epsilon Orionis), the middle star of Orion's Belt, which looks much dimmer than Rigel because its distance from Earth is about 1,300 light-years. A few other blue giants of spectral type O and B in Orion and neighboring constellations are even more luminous than Rigel and Alnilam. So Rigel gets its prominence in our skies by virtue of being closer than most of this winter region's other blue giants.

The many fiercely luminous blue giants in Orion and adjacent constellations are members of a vast congregation of young stars called the Orion Association. The existence of this association is, in fact, why people who casually glance at the night sky notice that the stars look brighter in winter. Most of these people think the greater brightness is due to the air being more clear in winter. In reality, cold nights are not always the clearest and many regions at midnorthern latitudes have their clearest nights in early autumn. The reason the stars look bright in winter is, along with a few chance close stars that happen to be bright, the Orion Association.

The Orion Association is an O-B association (see chapter 2). In fact, it is one of the nearest. And, in apparent magnitudes, it is the brightest of them all. When we look at these bright stars we are staring away from the center of our Milky Way galaxy and toward the center of the spiral arm in which we live—the aptly named Orion Arm. Not many millions of years ago a density wave must have passed through the dust and gas of this region, setting off a spectacular outburst of starbirth. That is why we see sparkling stellar sapphires strewn around this region of the heavens. The starbirth now is happening at a more modest rate but is certainly still occurring.

These new stars are bursting into radiance and beginning to burn out of the nebulas which gave them birth. The Orion Association is hundreds of light-years wide and deep but its heart lies in the region of the Belt and the Sword. There you will find not just giant stars but also star clusters, double and multiple star systems, and mostly dark nebulosity—which in some places, however, has been lit up brilliantly by young stars.

Clusters in the Orion Association? Well, Orion's Belt itself turns out to be part of a little-known cluster. On a very clear country night the naked eye shows a noticeable increase in the richness of stars within a few degrees of the Belt and the field is splendid in binoculars and wide-field telescopes.

What about double and multiple stars in the Orion Association? Several of the most important stars in Orion are interesting doubles. Rigel itself has a magnitude 6.8 companion 10″ from it, a delicate sight which usually requires a 6-inch telescope unless "seeing" is very good. The west star of the Belt, Mintaka (Delta Orionis), has a magnitude 6.3 star a full 53″ from it—a star not physically related to Mintaka but that nevertheless forms a very easy and lovely pair with it at low magnification in telescopes and even in binoculars. The eastern star of the Belt, Alnitak (Zeta Orionis) has a bright (magnitude 4.0) companion which is quite close to it—2.5″ away.

But suppose we demand systems with more than just two components? The Belt-Sword heart of the Orion Association has several of these several-star delights!

The most famous multiple-star system in Orion, as we'll see in a moment, is the Theta-1 Orionis system of at least six suns. But there are others. Sigma Orionis is the moderately bright naked-eye star about a degree southwest of the east star of the Belt, Alnitak. It is a widely separated quintuple star system, four of its stars (magnitudes 4.0, 10.3, 7.5, and 6.5) visible in amateur telesopes—with a faint triple star (triangle of magnitude 8.0, 8.5, and 9.0) about 3 1/2′ to the west. Then there is the bottom (most southerly) star in the Sword—Iota Orionis. Iota consists of a magnitude 2.8 primary with a 6.9 secondary 11′ away. The third star is a faint (11th-magnitude) one 50″

away at P.A. 103°. But what is easier and more beautiful to notice in the field of view in small telescopes is 8′ from Iota: another double (Struve 747), consisting of a 4.8 and a 5.7 star a full 36″ apart.

All of the above information on these double and multiple stars may read as a mere tangle of statistics to the beginner. But it translates in the telescope eyepiece into scene after scene of paired and congregated celestial jewels so varied in their respective brightnesses and arrangements that an observer could never grow tired of inspecting them.

The Horsehead and the Great Orion Nebula

Long-exposure images in visible and infrared light show much of the constellation Orion clothed in nebulosity. For a dark nebula to be visible in amateur telescopes it typically needs to be seen silhouetted against a bright nebula. The most famous example is an indentation about halfway down the eastern edge of a subtle strand of glowing gas that extends about a degree south from Alnitak (eastern star of the Belt). This tiny dark formation is known as the Horsehead Nebula, due to its striking resemblance to an interstellar knight chess piece on long-exposure photographs. Unfortunately, the Horsehead Nebula is a very difficult visual object. It has become a more common sight for amateurs with large telescopes in recent years by the use of hydrogen beta filters which make it easier to see. But it is possible to detect the Horsehead without filters. I've seen it a few times, using a 10-inch and a 13-inch telescope, and several great observers like Walter Scott Houston, Clyde Tombaugh, and Steve Albers have glimpsed it in 6-inch and smaller telescopes without filters.

There is another nebula in Orion, one which is as visually spectacular as the Horsehead is visually subtle. All that I have written about the Orion Association has in a sense led to this object, which is not the heart of darkness but rather the heart of light and beauty in the Orion complex. The object is M42, the Great Orion Nebula.

There are all the other deep-sky objects in the heavens, and then there is the Great Orion Nebula. No nebula shines so bright, and no object beyond the solar system displays such intricate detail. M42 would be very conspicuous to the naked eye if it did not so closely surround the wide double star Theta Orionis (magnitudes 4.6 and 4.8, 2′20″ apart), which interferes with the unaided view of the nebula

Figure 4.2. U.S. Naval Observatory image of M42, the Great Orion Nebula.

unless you have rather sharp vision. I referred a few pages back, however, to the way that binoculars or finderscopes show the nebula as an unmistakable puff of light which even in small telescopes expands to a star-sprinkled translucent fan, flower, seashell, or cupped hand of strong glow. As the size of the telescope used grows to about a 6-inch, the brightest part of M42, the Huygenian Region, begins to display a greenish hue and mottled surface. Even just a slightly larger telescope reveals an intricacy of branching wisps and whirls, patches of brightness and lanes of darkness, and enables one to trace almost a full circle and a full degree of nebula in really dark skies. More intense green and touches here and there of pink or pink-purple become visible to some observers. The long filament of light on the east side of the nebula becomes ever more prominent. So, too, does the great dark bay on the north end of the nebula known as the Fish's Mouth. It separates from the main mass of nebula a fine detached comma-shaped puff of nebula with its own Messier degination, M43. The Huygenian Region was named for Dutch astronomer Christiaan Huygens, who studied it in the seventeenth century. It begins in larger amateur telescopes to merit the account of it given in the nineteenth century by Sir John Herschel: "I know not how to describe it better than by comparing it to a curdling liquid, or a surface strewed over with flocks of wool, or to the breaking up of a mackerel sky when the clouds of which it consists begin to assume a cirrus appearance. . . . "

The heart and brightest part of the Great Orion Nebula contains a treasure which can be enjoyed in even just a good three-inch telescope. For, like four pieces of a broken comet nucleus within the brightest inner coma of a great comet, there shines in M42 the multiple star system Theta[1] Orionis, a quadrangle of stars which has been nick-named "the Trapezium." Components A, B, C, and D (named not in order of brightness, as is usual, but in order of increasing right ascension) shine at magnitudes 6.7, 7.9, 5.1, and 6.7 (though A and B vary in brightness). The sides of the Trapezium measure 9″, 19″, 13″, and 13″. Even a six-inch telescope begins to show subtle colors of the Trapezium stars, though they are probably illusory ones caused by the stars' contrast with the green background of the nebula. Larger amateur telescopes reveal a fifth and a sixth star in the Theta[1] system—both magnitude 11 and 4″ distant from, respectively, stars A and C. By the way, don't forget that Theta[2] is also visible, over in the east edge of the nebula, and is itself a double of 5.1 and 6.4 stars 54″ apart.

How bright is M42? When I first got a 13-inch Dobsonian tele-scope more than twenty years ago, I was shocked to see the nebula's light shining out of the eyepiece so brightly that it illuminated my hand with a slight greenish tinge. Deep-sky expert James Mullaney has used a mighty 30-inch refractor to project the nebula in color onto a white sheet of paper.

The Great Orion Nebula offers so much breathtaking detail that it is not unusual to find yourself studying it for a half-hour or even an hour at a time. But if we consider what this object is and what further wonders exist within it we will be even more awestruck.

In this region 30 light-years wide and about 1,300 light-years away, we are seeing a vast and ceaseless turmoil of starbirth seemingly frozen by the immensities of space and time. The nebula glows largely by fluorescence caused by the ultraviolet radiation of its hot young stars (the green glow is that of doubly ionized oxygen; the touches of pink are from the more familiar nebula emission of hydrogen). The Trapezium stars are thought to be no more than a few hundred thou-sand years old. Imaging at various wavelengths (including infrared and different radio wavelengths) by the Hubble Space Telescope and other instruments reveals a newly formed dust-wrapped cluster hidden beyond the visible Trapezium and other clutches of stars in the earliest stages of development. Dozens of *proplyds* (short for "proto-planetary disks"), solar systems in the making, have been pho-tographed in the nebula. There may be interstellar planets roaming the nebula as well.

One turns from M42 and Orion as if waking from a dream. But turn we must if we are to see other wonders of the January sky. In leaving, however, I hasten to recommend some additional sights for future visits to Orion. In the Belt-Sword region, there are many more sights we haven't discussed: the surprisingly easy nebula M78 (trace the Belt's length to its southeast end and then make a right—north-east—turn and go about one Belt length to find it); the overlooked nebula NGC 1977 around the magnitude 4.6 star 42 Orionis (the Sword star north of M42 and M43, and only about 4' from magnitude 5.2 star 45 Orionis); and the NGC 2024, the Flame Nebula, which needs good skies and a telescope of maybe a 10-inch aperture or more to see well so close to the east of Alnitak. Finally, don't think that the bright Orion stars and the sights of the Belt and Sword region is all there is to see in Orion. There is the shield and club of the Hunter to trace (the Orionid meteors of October come from the club and the shield contains side-by-side stars called Pi^3 and Pi^4 Orionis which are actually 26.2 and 1,100 light-years from Earth, respectively). The head of the Hunter is marked by a little naked-eye triangle of stars whose brightest member is Meissa (Lambda Orionis), a star which can be split into magnitude 3.6 and 5.5 components 4' apart. And W Orionis is a sixth-magnitude carbon star whose deep orange tint is a dramatic and welcome spot of warmth in Orion's fields of icy blue-white stars.

AURIGA

We travel north from Orion, crossing the horns of Taurus the Bull on our way to Auriga the Charioteer. Auriga's chief luminary, Capella, is ever so slightly brighter than Rigel. It lies almost exactly due north of Rigel but so far (54°) north of Rigel and near the zenith for people at midnorthern latitudes that this relation of the two is rarely noted. Auriga as a whole lies so high in the middle of January evenings that most of the world's naked-eye and binocular observers need to crane their necks to view it.

Capella is, at magnitude +0.07, the sixth brightest star in all the heavens, fourth brightest properly visible from midnorthern latitudes, and second brightest among the winter constellations. It shines with a subtle yellow hue. It may have the same general spectral type as our own Sun, but Capella is in reality a pair of suns that are each dozens of times more luminous than the Sun and maybe twelve and six times bigger, respectively. The duplicity (doubleness) of Capella was discovered spectroscopically, for these stars are closer together than Sun and Earth, whirling around each other in a 104-day period.

The name Capella means "she-goat" because this star is pictured as a mother goat being held by the mysterious figure of the Charioteer. Just a few degrees from Capella is a triangle of 3.0 to 4.0 magnitude stars known as "the Kids"—the baby goats of Capella.

The northern star of the Kids, the one nearest to Capella, is the eerie Epsilon Aurigae (sometimes called Ma'az or Al Ma'az). The light we see is from the primary star of this system, which must be one of the most luminous stars in our part of the galaxy. It normally appears as a magnitude 3.0 star, which from a distance of 3,000 light-years would make it comparable to Rigel in true brightness. But another source estimates Epsilon Aurigae at 7,800 light-years, which would give it a luminosity of –8.0, about 132,000 times greater than that of our Sun.

The mystery, however, is what kind of a companion star causes the regular but extremely infrequent and lengthy dimmings of Epsilon Aurigae A. The eclipses occur twenty-seven years apart! The star dims for months, then for about a year is at a minimum magnitude of 3.8 before brightening back for months. The original theory was that the cause of the eclipses was a very cool, very dim star much larger than any other known—big enough to fill our solar system out to the orbit of Uranus. The leading theory today is that the companion is surrounded by a huge disk of interstellar dust. The next eclipse of Epsilon Aurigae will occur in 2009 to 2011.

A little south and east of the center of the pentagon of stars formed by the major stars of Auriga (with the help of Beta Tauri) is an asterism and dim fuzzy patches which can be glimpsed with the naked eye in extremely dark skies. It is with fairly low magnification on a telescope, however, that the beautiful open clusters of middle Auriga are revealed in their full splendor. The best clusters in this part of Auriga are M36 and M38. M36 is dominated by B-type stars, like those in the Pleiades, but is about ten times more distant than the

Pleaides. M38 is bigger and its collection of B-type sparklers is dominated by a bright yellow giant. In telescopes, many observers see M38's main stars as forming a cross or Greek letter Pi (π). Yet although M36 and M38 are lovely, the most spectacular cluster in Auriga is M37, which shines at magnitude 5.6, contains about 150 stars, and rates as one of the few best open star clusters in all the heavens. Look for it just east of the southeast (Theta Aurigae to Beta Tauri) side of the Auriga pentagon.

About halfway between M38 and Iota Aurigae (southwest corner of the pentagon), we come to a remarkable reminder of Orion and the Orion Association. This is where you will find dim naked-eye variable star AE Aurigae. It is one of three "runaway stars" whose rapid motion when plotted back in time suggests that they were blasted out of the Orion Association a few million years ago. The other runaways are 53 Arietis and Mu Columbae. It has been suggested that all these stars were members of star systems in which another star went supernova, ejecting them away at great speed.

LEPUS

Lepus the Hare is one of the most underrated of constellations. It has usually been known just for M79. This globular cluster in Lepus is the only one easily visible to amateur astronomers across the vast span of sky from M2 and M15 in early autumn to M3 in early spring. You can locate M79 by taking the line from Alpha to Beta Leporis and extending it one more length to the south. Sometimes observers remember the almost ruddy carbon star R Leporis, also known as Hind's Crimson Star. R Leporis ranges in brightness up to about 5.5 but dims to as faint as 11.7, so it often requires a fairly large amateur telescope to see well. But what else of interest is there in Lepus?

One thing rarely appreciated is the brightness of the major stars of Lepus. Alpha and Beta Leporis (Arneb and Nihal) shine at 2.6 and 2.8 (hardly dimmer than a few of the main stars of Orion), and six other stars of Lepus are brighter than 4.0. Furthermore, Arneb is in reality a tremendously bright star—a member of the Orion Association almost as luminous as two of the Belt of Orion stars, which lie about 18° due north of it. Arneb shines more than an hour of R.A. almost due west of Sirius. The main pattern of Lepus thus seems to be very conspicuously located, at a right angle to Orion and Sirius. But that is just the problem: The brightest constellation and brightest star are unfair competition, leaving Lepus usually overlooked, despite its Beta Leporis shining at 2.6 and six other stars that are brighter than 4.0.

Double stars have gotten too little attention in recent decades but Gamma Leporis will fire your interest. Even binoculars can split this pair of stars 96″ apart—but use a small telescope at low power to best see the lovely yellow and orange hues of the 3rd-magnitude primary and 6th-magnitude secondary.

COLUMBA

South of Lepus is Columba the Dove. Columba has a few moderately bright stars but its southern declination means they are dimmed by atmospheric absorption for most observers at midnorthern latitudes. Columba also has a few fairly interesting deep-sky objects—but only a few. Among these are 5.1-magnitude Mu Columbae, one of those runaway refugees from the Orion Association.

There is, however, a lonely point in northeast Columba, near its border with the constellation Canis Major, which has a very special meaning for residents of our solar system. It is the Antapex of the Sun's Way. This is the spot in the heavens from which our Sun and its system of planets is departing. When we look here we are, in a very real sense, looking out the back window of the spaceship which is our solar system as it pursues a 200-million-plus-year orbit around the center of our galaxy. There is an older, very beautiful term for the solar antapex: the Sun's Quit.

THE LORE OF JANUARY'S CONSTELLATIONS

The constellation Orion has been identified with some of the most powerful gods and heroes of ancient cultures—for instance, with the Egyptian Osiris. Yet some cultures seem to associate this show of flashy brilliance with legendary figures who were particularly vain about their physical beauty, power, or prowess.

The latter is certainly the case with the ancient Greek character Orion. Orion was said to be the handsomest of mortal men (though he was part god, being son of the sea god, Poseidon, who the Romans called Neptune). Orion was so tall that when his feet were on the sea bottom, his head was above the waves—or, according to another account, he inherited from his father the ability to actually walk across the surface of the sea. (Either way we can see this as a beautiful mythologizing of the experience of seeing the constellation looming large at its rising or setting on a sea horizon.)

Orion was violent and lustful, and among the many maidens he pursued with lascivious intent were the Pleaides sisters—indeed, storytellers say, look up on a winter night and you can see that his pursuit of them continues even now that he is a constellation and they a star cluster. Another story tells of Orion's death and a scorpion he battled. Hera, queen of the gods, hearing of Orion's boastfulness sent a giant scorpion to sting and kill him. After their deaths, both hunter and scorpion were converted to constellations and their conflict supposedly continues even in the heavens. Skywatchers can indeed note that when the constellation Scorpius the Scorpion makes its first appearance in the east after nightfall in late spring, Orion has just set, having fled from the night sky.

Legends have arisen about the Orion constellation and some of the neighboring star figures in the sky. Canis Major the Big Dog and Canis Minor the Little Dog are often imagined to be the hounds of Orion the Hunter. Lepus the Hare is fleeing from the Hunter or perhaps just from his dogs. Orion has also been imagined to be confronting the charging Taurus the Bull, holding his lionskin shield west toward Taurus while he uplifts his club in his arm northeast of Betelgeuse.

Auriga the Charioteer is a mysterious figure with his chariot or wagon and goats, but is sometimes identified in Greek myth with Erichthonius, supposed inventor of the chariot. Lepus the Hare can stand for any of the many hares and rabbits of myth, including those associated with the Moon (according to some storytellers, the markings on the Moon are in the shape of a hare who leaped all the way up there). Columba the Dove was invented in 1572 by Petrus Plancius, who wanted the nearby ancient constellation Argo Navis (the Ship Argo) renamed Noah's Ark and intended Columba to represent the dove which Noah sent out to search for dry land. But we might also imagine it to be the dove which the Argonauts sent out to see if it could get through the Clashing Rocks (it did, just barely, assuring them that they would have time to row the Argo through before the rocks crashed back together).

SPECIAL SIGHTS OF JANUARY

Orion, Auriga, Lepus, and Columba are not the only sights we can seek to observe every year in the January night sky. About two days after celebrating New Year's Day, it is time to take a look for the peak of the Quadrantid meteor shower.

The radiant of the Quadrantis at their peak is located in the northeast part of the constellation Bootes. The shower takes its name from the abandoned constellation Quadrans Muralis (the Wall Quadrant) whose stars now belong to the constellations Bootes, Hercules, and Draco. Viewers around 40° N latitude will find this radiant on the north horizon around 8:30 P.M., low until midnight, and highest (about 60° high in the northeast) as dawn comes.

The peak of the Quadrantids falls on January 3 or 4. If the Quads were like most annual meteor showers, the best numbers of them you'd see would tend to be when the radiant was highest on the best night. Instead, the Quadrantid meteor stream is so narrow that the shower is above quarter strength for only about fourteen hours and the really high numbers occur for only a few hours. Consequently, the exact time of the maximum is very important. In years when it occurs late in the daytime at your location, the numbers of Quads you see at the preceding and following dawns might not be very impressive (maybe only five or ten per hour even in a clear, moonless country sky). On the other hand, in years when the maximum takes place not long before morning twilight at your longitude and latitude, the Quadrantids can be your strongest meteor shower of the year. You may then see dozens or even well in excess of one hundred of them in your best hour of observing.

Check with your astronomy magazine or meteor organization Web site of choice each year to learn when the maximum of the Quadrantis is expected and which night or morning it will be best viewed at your location. What you will see are moderately bright, moderately fast meteors which have a reputation for often being blue and sometimes leaving "fine long spreading silver trains."

JANUARY'S OBSERVING EXPERIENCE

Those hours spent looking for Quadrantids before dawn are often the coldest of the year. Even if you confine your observing to the evening hours, the overwhelming practical need for January observing is staying warm.

It's important to remember that it is much harder to stay warm during a mostly stationary activity like observing than it is while skiing or walking. You simply cannot enjoy even the glory of the Orion Nebula in a telescope when your teeth are chattering, your body shivering, and your hands and feet stinging.

Many of us know the basics of dressing for the cold. But it is smart to review them and to add a few lesser-known tips, because astronomical observing is an extreme cold-weather challenge.

Radiational cooling is at its maximum from anything that is exposed to the sky on a clear night. You must keep your heat from draining away into space. A hat or hood is essential, for the body loses much of its heat through the head. Hands and feet are the farthest places to which your blood is pumped, so they are the first that your body will cut down the flow to if your body has to conserve heat for the rest of you. In very cold weather it may be necessary to wear thick outer gloves which unfortunately don't allow much dexterity. In that case, you can also wear thinner gloves or mittens underneath them so that you can briefly take off the heavy, outer glove when you need to have the finesse of your fingers to focus the telescope or change an eyepiece. Layering clothes is a good idea, so long as you don't wear such thick and multiple layers of socks that you can barely squeeze your feet into your shoes or boots. Doing so could cut off the blood flow to the feet and make things even worse.

The key thing to remember is that you must create as great a thickness of "dead air space" as possible between your body and the outside air. Although the exact meaning of "dead air space" may be debatable, the point here to appreciate is that the material with which your coat is made is of only secondary importance. The only reason that coats made of material like down are prized is because they can provide you a given thickness with less weight to burden you.

You may have heard that drinking alcohol constricts blood vessels and therefore is a very bad idea if you are planning to be out in the cold. But did you know that it is also important to drink sufficient quantities of other fluids—preferably water—if you want to keep from getting cold? The body can easily get dehydrated in the cold (and in the usually dry air of a clear winter night) and it is less able to maintain itself against the chill when it is not properly hydrated.

Do these special preparations for facing the cold sound like annoyances? Perhaps you won't really need most of them? Don't bet on it. If you don't follow these guidelines you will find yourself absolutely miserable when you start shivering, psychologically and physically unable to enjoy the treasures of winter starlight.

On the other hand, if you do dress properly, few nature experiences are more thrilling than standing underneath a clear, starlit winter sky. There is a solitude and a pristine, seemingly virtuous quality to the winter night. The only sound you may hear is the crunch of snow, ice, or even ground under your feet as you walk, or the occasional crack of a branch snapping in the freeze, or the resonant hooting of great horned owls sounding near even from a mile away. Then silence resumes. And silence on both earth and in the sky makes starlight all the more eloquent.

Chapter 5

February

[Constellations covered: Canis Major, Canis Minor, Puppis, Monoceros, Gemini]

If you live in a northern land, February is probably your snowiest month. This is certainly the case almost everywhere in the U.S., for the storm-guiding jetstream reaches its most southerly position and air masses are, on the average, almost as cold as in January. Snow-storms mean cloudy skies but unfortunately even after the storms there are problems getting around in the snow, especially with a tele-scope. Another problem that few beginners would foresee occurs if there is any extraneous artificial lighting in the environment: dark adaptation is compromised by light reflected off the snow on the ground.

Although snow on the ground can be a problem for skywatchers in February, it can also add much beauty to a landscape still ruled over by many of January's brilliant stars and constellations. In the middle of February evenings, Orion has moved into the southwest and is beginning to decline. But this is when Sirius, brightest of all stars, reaches its culmination in the south in Canis Major the Big Dog. Well upper left from Sirius is Procyon, the chief light of Canis Minor the Little Dog, and north of Procyon shine Pollux and Castor, the twin bright stars of Gemini the Twins. In February, Pollux and Castor hang almost overhead at midnorthern latitudes around midevening.

February evenings are also the time to use a telescope to search for the many exotic star clusters and nebulas in the dim constellation of Monoceros the Unicorn. It's a time to look for the subtle winter Milky Way's soft sash of light, and to try to glimpse on a dark country evening one of the sky's other elusive glows, the zodiacal light.

Our all-sky map for February shows that Perseus is still high in the west-northwest, with long Andromeda trailing below it to the horizon. Facing north, Cassiopeia is in the northwest about as high as Polaris and the Big Dipper is in the northeast and a little higher than Polaris. In the east, Leo has leaped to almost halfway up the heavens, while in the southeast the front half of the water serpent, Hydra, stretches forward. Dim Cancer the Crab is high up between Leo and Gemini, while the blazing star Arcturus is just minutes away from rising in the east-northeast.

Our close attention is not on Perseus or the Big Dipper or Leo this month, though. It is with Canis Major, Canis Minor, Monoceros, and Gemini. And we begin our study with the Dog Star of Canis Major: peerless Sirius.

SIRIUS

Sirius is the most brilliant of all night's stars. The competition isn't even close. Sirius shines about twice as bright as the second-brightest star Canopus, about four times brighter than Capella and Rigel, and nine times brighter than a standard first-magnitude star. Sirius burns at an apparent magnitude of –1.44. Of course, the brightest planets can shine brighter than this. But they do not twinkle. In the December chapter, I mentioned the beauty in this twinkling of Sirius, including the color changes, and how the twinkling is most prevalent when Sirius is low in the sky—an hour or two after its rising or before its setting. But for viewers around 40° N latitude, Sirius never climbs more than 32° above the horizon, even when highest in the middle of February evenings (and it is lower in your sky if you live farther north than this). Also, there are some nights when the atmosphere is much more turbulent. At those times, the scintillation of Sirius can be prominent even when it is highest. On the other hand, if you get a night of steady images, try to judge the true color of Sirius. Some people see its true color as whitish, but I have found that I and perhaps most people detect a lovely touch of blue in its radiance.

Sirius is known as "the Dog Star" due to its marking the head (some say heart) of Canis Major the Big Dog. But what of the name "Sirius" itself? It is pronounced like the English word "serious" (leading to many groan-inducing puns). The ancient Greek meaning of this name appears to be "sparkling" or "scorching," appropriate for a star of such blazing brilliance. The latter translation even suggests light so great that it is accompanied by heat. And that is apt, considering the connection between Sirius the Dog Star and the "dog days" of summer.

How did the expression *dog days* come into being?

Since Sirius is highest at night in winter, we know that it must be highest in the day in summer. Many ancient people realized this and some could figure out that Sirius is in the same general region of the sky as the Sun during the hottest days of summer. This fact led to a classical Greek and Roman belief that those hot and sultry days were caused by the heat of Sirius being added to that of the Sun. Thus they were called "the dog days" or "canicular days." Different lore traditions suggest different exact dates for the dog days but perhaps the most widely followed tradition today places them from July 3 to August 11.

In modern times, August 11 is indeed about the date when people at 40° N might get a first glimpse of Sirius returning to visibility after being hidden by the Sun's glare for days, weeks, or even a few months. The star is glimpsed low in the southeast during bright morning twilight. Astronomers have a name for the first visible rising of any celestial object which has been invisible for a while due to proximity to the Sun. They call it the *heliacal rising* (Helios was the original Greek god of the Sun). The heliacal rising of Sirius was once so important it was greeted with New Year celebrations in ancient Egypt, for the heliacal rising of Sothis (Sirius) signaled the coming of the life-giving annual flood of the Nile. It was this flood that made fertile the soil near the river's banks.

THE DOG STAR AND THE PUP

If Sirius was merely spectacular in appearance, that would be enough, but this star is also interesting as a sun for several reasons.

First of all, Sirius is the closest star easily visible to the naked eye for most of the world's population (only if you live in the southernmost U.S. or farther south can you glimpse the closest star or star system of all, Alpha Centauri). Sirius is located just 8.6 light-years from Earth. (By the way, near and far are relative: Sirius may be one of the closest of night's stars but it is more than half a million times farther than our day star, the Sun.) It is fun to point out to eight- and nine-year old children that the beautiful light they are seeing from Sirius tonight actually left the star around the time they were born. (On a February evening, there are other bright stars to use for other ages of people, too: Procyon for eleven- or twelve-year-olds; Pollux if you're 34; Capella if you're 42; Castor if you're 52; Aldebaran if you're 65; and Regulus if you're 78.)

Sirius may derive most of its brightness in our sky from its closeness, but it is a spectral type A star not only far hotter but also far more luminous than our Sun. It has more than twenty times the luminosity and more than twice the mass, packed into well over 1 1/2 times the diameter of our Sun. It is easily the most luminous star in our solar neighborhood until we get out to just over 25 light-years from Earth and encounter Vega. Interestingly, winter's blue-white Sirius is not far from the Antapex of the Sun's Way (see January chapter) and summer's blue-white Vega is not far from the Apex of the Sun's Way. Thus we see similar stars behind and before us: Sirius out the back window of our traveling solar system, and Vega out the front or windshield.

There is a secret about Sirius which makes it even more interesting. The secret was not suspected until the nineteenth century when F. W. Bessel discovered that there was a wavy component to Sirius's motion through space. This anomaly could best be explained by the existence of a companion star tugging on Sirius. However, it was not until 1862, almost two decades after the conclusion of Bessel's study, that Alvan G. Clark, using his superb 18 1/2-inch refractor, became the first person to sight the companion of Sirius. The companion, which became known as Sirius B or, more playfully, "the Pup," is a magnitude 8 1/2 star which is difficult to see because of its proximity to Sirius. Sirius B is always a challenging object for amateur astronomers but is especially difficult during the many years around the time it is closest to Sirius in our sky. The Pup pursues a fifty-year orbit around Sirius, its apparent distance from the main star varying from as little as 3″ to as much as 11 1/2″. The companion was

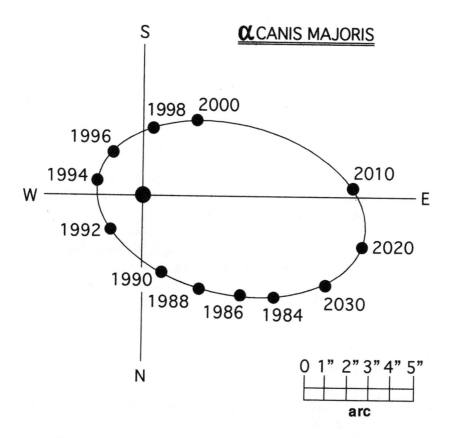

Figure 5.1. Changing position of Sirius B (smaller dot) relative to Sirius in our sky.

last at its widest separation from Sirius in 1975 and will be again in 2025—so its visibility is improving in the opening years of the twenty-first century.

It is interesting that the brightest star has a companion almost exactly 10 magnitudes or 10,000 times dimmer. But the story gets more exciting. Astronomers were able to calculate from the Pup's amount of tug on the motion of Sirius how much mass the companion has. They found that its mass was similar to that of our Sun. But for this to be true of such a dim star could only mean one of two things: either Sirius B had a very low surface brightness and resultant low luminosity or it was a hot but incredibly small star. Which was true? The spectrum of Sirius B indicated it was almost as hot as Sirius itself. So the Pup had to be a star of astonishingly small size: perhaps only about 2 percent as wide as the Sun—that is, only about twice as wide as the Earth. Yet if as much mass as our Sun is compressed into a body only twice as wide as Earth, the density of material in Sirius B has to be tremendously great. A typical cubic inch of Sirius B must weigh about two tons! And the only way this can be possible is if the star is largely composed of atomic nuclei and loose electrons that are packed together—what scientists call *degenerate matter*.

In short, Sirius B turned out to be the first clear-cut example of one of the most important types of stars—a white dwarf. As we saw in chapter 2, the fate of the Sun and many other stars is to become a white dwarf as a last stage in their life as a luminous object. Sirius B is closer to Earth and has greater apparent brightness than any other white dwarf. (It is not the easiest white dwarf to observe, though. That distinction falls to Omicron Eridani C—discussed in our December chapter—because it is not close to a tremendously brighter primary star.)

The discovery of Sirius B has given rise to some modern-day legends. For instance, it has provided an appealing, but highly improbable, explanation for why a number of ancient writers refer to the color of Sirius as if it were reddish. Before Sirius B became a white dwarf it was a red giant. Could Sirius B have been a red giant just two thousand years ago so that its light blended with that of Sirius to make the naked-eye point of light look somewhat ruddy? Recent research suggests that the Pup could not have been a red giant so astronomically recently. The attribution of ruddy color to Sirius by some ancient writers is far more likely to be a result of seeing red from dimming by haze or in the prominent twinkling of this often low and very bright star.

Much stranger is the claim about Sirius B and a primitive people called the Dogon. The Dogon live in Mali in Africa. When they were first studied by anthopologists in the 1930s, the Dogon seemed to have a tradition about a companion star which went around Sirius in fifty years and contained material so heavy that all beings on Earth combined cannot lift it! The Dogon also knew that the Earth goes around the Sun. They even knew that Jupiter has four big moons, and that there is a kind of ring around Saturn—despite the fact they had no telescopes. Could they have learned these facts from extraterrestrial visitors? The stupendously more likely possibility is that the Dogon had been visited by some missionary, explorer, or scientist with astronomical knowledge sometime between the discovery of Sirius B in 1862 and the first visit by anthropologists in the 1930s. If the Dogon had long worshipped the star Sirius as a deity, then perhaps a visitor from Western civilization may have wanted to give them this marvelous new information about their deity.

The Other Wonders of Canis Major

The constellation in which the brightest star resides is one of the brightest and most interesting in the heavens. Yet it is often overlooked by amateur astronomers who are distracted from its other wonders by the greatness of Sirius.

One Canis Major object nevertheless well-known is the open star cluster M41. In this case, Sirius may actually help draw some attention to the cluster, for M41 is located only about 4° south of the brilliant star. This large and rather rich cluster can be glimpsed as a naked-eye spot of glow under very good observing conditions. As a matter of fact, M41 was apparently mentioned as early as 325 B.C.E.—by Aristotle, no less. It is one the best of all clusters in binoculars and in telescopes at low power. The cluster's brightest star is a reddish one near the center which shines at magnitude 6.9. The overall magnitude of M41 has been estimated at 4.5.

Another cluster in Canis Major is far less famous than M41 but lovely in its own right. This is NGC 2362, also known as the Tau Canis Majoris Cluster, because it surrounds magnitude 4.4 Tau Canis Majoris. This cluster may lie a little more than 5,000 light-years from Earth and if Tau is a cluster member (as is now thought), then this blue giant must be one of the most luminous stars in our part of the galaxy. Tau overwhelms the tightly crowded gems of the rest of the cluster in the view through binoculars but the lesser jewels of this compact gathering are brought out by a telescope. NGC 2362 is less than a million years old—one of the youngest clusters known.

Just a few degrees north of NGC 2362 is the finest color-contrast double star of winter. This very little known wonder has been listed in catalog and on star atlases as h3945 (Herschel 3945) but is also known as 145 Canis Majoris. It appears to the naked eye as a magnitude 4.5 star. Even the lowest telescopic power splits the point into a 4.8 and 6.8 star that are 27″ apart. This pair has been compared to the most famous of all colored doubles, the summer star Albireo. There is a temptation to draw more attention to h3945 by calling it "the Albireo of Winter." In reality, even though the brighter star of each of these doubles has the same spectral class—K3—the dimmer member of h3945 is much cooler than Albireo B. I find its colors quite different than Albireo's gold and blue. (One astronomy writer has called the colors of h3945 "fiery red and greenish blue"—see what you think.) By the way, although the brighter star of h3945 has the same spectral class as the Albireo primary, it is a much mightier star, shining to us from about 2,000 light-years away. The dimmer member of h3945 is not physically related to the brighter, being about 250 light-years from Earth.

The double h3945 is located just a few degrees east of Omicron^2 Canis Majoris, one of at least five stars in Canis Major which may have luminosities as great or greater than Rigel.

Some of the stars of Canis Major have a far greater apparent brightness than most observers realize—for everything seems pale compared to Sirius. Beta Canis Majoris (Murzim), several degrees west of Sirius, burns at magnitude 2.0. But note especially what I sometimes call "the South Canis Triangle"—the compact triangle of Delta, Eta, and Epsilon Canis Majoris (Wezen, Aludra, and Adhara) at the southern end of the constellation's main pattern. The magnitudes of these three are 1.8, 2.4, and 1.5! Although this triangle is considerably less compact than Orion's Belt, the combined brightness of its stars is only marginally dimmer than the Belt's. Of course, the South Canis Triangle is almost 30° farther south than the Belt. Therefore, observers at 40° N latitude should try to admire the triangle when it is highest (for instance, in the middle of February evenings), for even then they see its stars dimmed by 0.3 or 0.4 of a magnitude by atmospheric extinction.

Adhara is thus slightly dimmed for most of the world's population, so they see it appear a little dimmer than Gemini's famous Castor. In reality, the apparent magnitude of Adhara is almost a tenth of a magnitude brighter than Castor. The traditional list of 1st-magnitude stars ends with Regulus, whose brightness is 1.36. But technically we could consider any star with a brightness between 0.5 and 1.5 as being 1st-magnitude—and Adhara shines at 1.50. Nevertheless, it will probably continue to be considered the brightest of the 2nd-magnitude stars rather than the dimmest of the 1st-magnitude. Incidentally, Adhara has a magnitude 7.8 companion just 8" from it. It can be considered a warm-up exercise for finding Sirius B. If you can't detect the companion of Adhara, you needn't even bother looking for the very much harder to see Sirius B.

PUPPIS LATER, CANOPUS NOW

If you are already looking as far south as Adhara and the southern Canis triangle and you have a clear view to the triangle's left and below it, you should be noticing a lot of fairly bright stars. Which constellation do these belong to? The answer is Puppis, the Poop Deck— of the much larger former constellation Argo the Ship (Argo was so huge and had so many stars that it had to be broken into several separate constellations that represent parts of the ship).

February is a good month for people at midnorthern latitudes to use middle and late evening to observe Puppis at its highest. There are reasons, however, why it is more useful for us to discuss Puppis and the other very southerly constellations formed from the old ship Argo in the March chapter. You might want to refer to that chapter and get a jump on Puppis while sunsets are still early and Puppis is higher.

There is, though, one prominent star of old Argo which is farthest west and needs to be looked for by people at mid-northern latitudes no later than February. I am referring to the brilliant light which leads the entire Ship across the sky—the second brightest of all stars, Canopus.

Canopus shines in Carina the Keel. It is the most northerly of the bright stars that are traditionally considered south circumpolar. Yet it is too far south to be seen from above about 37° N latitude. In fact, even from considerably farther south—say, southern Florida or Texas, it is low enough for atmospheric absorption to cut its brightness in half. Thus even for the hour or so that Canopus is near its peak height in these places, it never looks brighter than Rigel or Capella. The undimmed Canopus shines at –0.62, only about half as bright as Sirius but still markedly brighter than any other stars. Remarkably, Canopus lies roughly due south of Sirius (about 35° of declination south of Sirius), so when the latter is at its highest, so, too, is Canopus. Canopus is reputed to have a bit of a yellow tint but this is probably just due to many observers seeing it low and reddened by haze and thick atmosphere. Where it is visible high in the sky, Canopus is white.

Canopus is a mighty sun. It is the supergiant nearest to Earth, lying 310 light-years away. Its luminosity is almost fifteen thousand times that of our own sun.

PROCYON, CANIS MINOR, AND THE WINTER MILKY WAYS

Just as Sirius tends to overshadow the other wonders of Canis Major, so, too, does it overshadow the other Dog Star—Procyon.

The name Procyon means "before the dog" due to its role in announcing the rising of Sirius. Interestingly, although Procyon rises before Sirius for viewers around 40° N latitude, it is almost a whole hour of right ascension east of Sirius. The key to its earlier visibility is its more northerly declination. Procyon is located at +5°13′, about 21 1/2° of declination farther north than Sirius. Whereas Sirius starts setting in bright evening twilight in April, Procyon is easily visible into May. (By the way, Procyon is almost due east of Betelgeuse—which it usually outshines a little.)

Procyon has many interesting characteristics, but in almost every one of them it is exceeded by Sirius. Procyon glows at 0.40, making it the eighth brightest star of all and sixth brightest from midnorthern latitudes. But of course Sirius is the brightest star and outshines Procyon by almost two magnitudes. Procyon is the outstandingly bright star of Canis Minor. But Sirius is the outstandingly bright star of the bigger brighter Canis Major. Procyon is the second-closest bright star properly visible from midnorthern latitudes, at 11.4 light-years away. But Sirius is the closest visible from those latitudes, at 8.6 light-years. And even though Procyon is considerably farther away than Sirius, its true brightness (luminosity) is still considerably less than that of Sirius.

When it comes to the nature of the companions of Sirius and Procyon, there are two ways of looking at the situation. Procyon B is much more difficult to observe than even Sirius B—so difficult it has never been seen in an amateur telescope. Procyon B is simply too faint and close to its bright primary (magnitude 10.8 with a separation varying between 2″ to 5″ over the course of forty years) to see with anything less than a huge telescope under excellent conditions. On the other hand, we can't help being impressed by the fact that Procyon B is probably an even smaller white dwarf then Sirius B.

Procyon is less massive than Sirius but probably a little bigger. It is an F5 star, thus cooler than Sirius but hotter than the Sun. Can you see a very slight hint of yellow in the white of Procyon? Try to estimate its color with the naked eye, binoculars, and a small telescope.

We discovered that there was a great deal more to Canis Major than just Sirius. The same cannot be said of Canis Minor and Procyon. Canis Minor covers a small area of the sky with few deep-sky objects of interest. The pattern of the Little Dog is usually shown as just a single line connecting Procyon with Beta Canis Minoris, the modestly bright (magnitude 2.9) Gomeisa.

Although Sirius and Procyon are not too far from each other in either space or sky and are linked in the human imagination by both being Dog Stars, the two are separated by something in the sky: the winter Milky Way.

There are even a few legends about the luminous band of the Milky Way separating characters represented by Sirius and Procyon. But these have not nearly reached the level of fame and richness achieved by the Far East's great myth of the spinning damsel and cowherd (the stars Vega and Altair) separated by the summer Milky Way (see the July chapter, under Aquila). That's not surprising considering how much dimmer this part of the Milky Way band is that passes through the traditional winter constellations—down from Auriga and Gemini and between Canis Major and Canis Minor on its way to Puppis and Vela. Yet there is a satisfaction when you see the winter Milky Way between the Dogs at all, because you then know you have a dark and clear sky.

Incidentally, one of the reasons we do not see as much glow from star-clouds in this winter Milky Way as in the summer's is very interesting. It is because the winter Milky Way stars between the Dogs lie within our own spiral arm of the galaxy so that many are close enough to be seen indiviudally with the naked eye under very good sky conditions.

MONOCEROS

The sky between Canis Major and Canis Minor is owned by a constellation with few bright stars but many dim ones—and amazing deep-sky treasures for telescopes. The constellation is supposed to represent one of the most famous types of imaginary creatures—the unicorn. The name of the constellation, Monoceros, means "unicorn." Or, if we translate further, both Monoceros and unicorn mean "one-horn."

Almost all of Monoceros lies within a triangle of brilliant stars that shine in neighboring constellations. This asterism (pattern of stars that is not an official constellation) is sometimes called the Winter Triangle. It is formed by the first-, sixth-, and seventh-brightest stars visible from 40° N latitude: Sirius, Procyon, and Betelgeuse. The reason why this pattern has not taken on the popularity of the Summer Triangle (Vega, Altair, and Deneb) is that there are other brilliant stars nearby which naturally blend with it—Betelgeuse especially is locked into the more compact and compelling pattern of Orion. Nevertheless, the Winter Triangle is useful for finding your way to some of the exciting telescopic sights of Monoceros.

Beta Monocerotis and M50

The brightest star of the Unicorn is magnitude 3.7 Beta Monocerotis, which many observers feel is the most beautiful triple star in the heavens. You can find Beta (and a few degrees to its right, Gamma) Monocerotis about a third of the way along the line from Sirius to Betelgeuse. A 6-inch or larger telescope on a steady night turns Beta into a thrilling acute triangle of stars with magnitude 4.7, 5.2, and 6.1. Star B (magnitude 5.2) is 7″ from the brightest star, and star C (6.1) is 10″ from the brightest star. B and C are only 3″ apart. Amazingly, all three stars are of the same spectral class—B3. They should all look white or bluish white yet some observers see yellow in the stars and/or some differences in the tints. The duplicity—or in this case, triplicity—of Beta Monocereotis was discovered by Sir William Herschel in 1781 (the year he also discovered the planet Uranus). The Beta Monocerotis system lies about 650 light-years from Earth.

A little more than one-third of the way from Sirius to Procyon is one of the finest of winter's many open clusters: M50. At magnitude 5.9, M50 is capable of being glimpsed with the naked eye under excellent conditions, though there is a lack of good naked-eye stars nearby to use as guides. More than one observer has noticed that some of the brighter stars of the cluster form a heart-shaped outline. Some also notice that the cluster does contain one star which is noticeably ruddy. M50 has many dozens of faint stars and so benefits from viewing with a medium-size or larger amateur telescope.

The clusters M46 and M47 are in Puppis, not Monoceros. But they are only a few average binocular fields-of-view southeast of M50. It is tempting to discuss them and Puppis in February, even though they have not yet reached their highest at midevening, because March nightfalls come later and eat into observing time. But if you wish to leap ahead to these objects and Puppis—going out a little later on February evenings to see them—just consult the coverage of them in the March chapter.

The other telescopic showpieces in Monoceros are on the opposite end of the constellation from M50. As a matter of fact, they are located on or near a line of three naked-eye stars, which I like to think of as the very horn of the Unicorn.

Deep-sky Objects of the Unicorn Horn

This short line of three evenly spaced stars is formed (from southwest to northeast) by Epsilon, 13, and 15 Monocerotis—all about magnitude 4 1/2. The middle star, 13 Monocerotis, lies almost precisely between Betelgeuse and Procyon, about one third of the way from the former to the latter. Once you find the Unicorn horn, you may notice that there is an only slightly dimmer star—or is it too fuzzy to be a star?—just 2 1/2° east (left) of the Epsilon Monocerotis, the bottom star of the horn. What you are seeing—pretty easily if your sky is quite dark—is a magnitude 4.8 star cluster which few amateur astronomers realize is such a plain naked-eye object. The cluster is NGC 2244 (that's an easy number to remember!). It is famous for being enwreathed by one of the most photographically spectacular nebulas in all the heavens—the Rosette Nebula.

NGC 2244 is a big lovely sight in even small telescopes and sports as brightest member the yellow, sixth-magnitude star 12 Monocerotis. But the much larger Rosette Nebula is a difficult object to see without a nebula filter. What you can expect on a very clear, dark night with a 6-inch telescope at low power is elusive averted-vision hints of the "petals" of the Rosette here and there around the cluster. Under superb conditions, large binoculars or small rich-field telescopes can show a soft blur around the cluster which is the entire nebula, more than 1° across. The Rosette is usually listed as NGC 2237 but originally was given this and three other NGC numbers to describe different bright parts of the nebula.

Now consider the most northerly of the three stars of the horn—the tip or end of the horn. This is 15 Monocerotis, a very hot O7 sun interesting in its own right. But it is also the brightest star of NGC 2264, a bright cluster shaped like a long triangle or evergreen tree and therefore named the Christmas Tree Cluster. The tree is almost 1/2° long and visible in binoculars. 15 Monocerotis (also known as

S Monocerotis—a variable star designation, for its brightness changes slightly) is the base or trunk of the Christmas Tree. The star at the peak of the tree is almost three maganitudes dimmer than 15 Monocerotis but slightly brighter than any of the other stars in the cluster. Just beyond this "top" (actually it's the southern end, but it looks like the top in the inverted view of most telescopes) is another of the sky's greatest photographic marvels, the Cone Nebula. It is a cone of dark gas and dust outlined by glowing nebulosity. Only a slight hint of it can be glimpsed through large amateur telescopes under superb sky conditions. But the view of the Cone Nebula on photographs prompted deep-sky author Robert Burnham Jr. to say that if primitive people could see such a view they would call it "the Throne of God."

There is a third famous Monoceros nebula only about 1° southwest of the Cone Nebula and the tip of the Christmas Tree Cluster. This is NGC 2261, Hubble's Variable Nebula. Unlike the Cone and Rosette Nebulas, this one is of such high surface brightness as to be visible in even small telescopes. NGC 2261 surrounds the variable star R Monocerotis. The nebula measures only about 2' by 1'. It impresses everyone as looking cometlike with a brighter, seemingly denser "head" from which extends a short "tail." The star looks like a cometary nucleus in the head—when the star can be seen, for sometimes this magnitude 8 to 10 point of light is overwhelmed by the high surface brightness of the nebula. What's amazing is that the nebula itself is variable in brightness and its changes do not seem related to the fluctuations of the star's brightness. The nebula's changes in brightness and shape are often very swift—sometimes even noticeable in hours—and irregular as to timing. The best theory, as Robert Burnham Jr. says, seems to be that the nebula's variations are caused by "changing light conditions and moving shadows cast on the cloud by dark masses drifting near the illuminating star"[2]—what an eerie scenario!

In 1949, Hubble's Variable Nebula became the first object photographed by the famous two hundred-inch Hale telescope at Mt. Palomar, California, which for decades remained the world's largest telescope.

GEMINI

North of Canis Minor and Monoceros is one of the most famous constellations, Gemini the Twins. This zodiac constellation is appropriately named, for its brightest stars are two very bright ones less than 1/2 magnitude different in brightness and only 4 1/2° apart. At first glance these two stars do seem almost twins. And they are named for the most famous twins in history, brothers in Greek and Roman mythology: Castor and Pollux.

Before we consider any mythology, however, we'll study the stars Castor and Pollux—what we can see of them and what astronomers have learned about their nature.

Pollux and Castor

Pollux and Castor turn out to be not as similar as they seem.

Let's focus on their differences.

Pollux beams at magnitude 1.14, the seventeenth brightest star of all, while Castor glows at 1.58, the twenty-third brightest. Not very different. But it is Castor that is the brighter star in reality. It exceeds Pollux in absolute magnitude, its true brightness being 0.6 to Pollux's 1.1. The distance to Pollux is only 33.5 light-years, to Castor's 52.

The next differences are major. Pollux shines with a definite orange tint (you may need binoculars to see this well) in contrast to the hotter, white Castor. Pollux is a K class star, Castor a much hotter A class star. Pollux is, as far as we can tell, a single star (though there are a few faint optical—merely line-of-sight—companions of it when you look in the telescope). And Castor? Castor is a system of at least six stars related to each other in space!

Castor A and B are spectacular in amateur telescopes, a close-together pair shining at 1.9 and 2.9. In the 1960s and 1970s they were too close together—just 1.9″ apart in 1970—to see in small telescopes. But they have widened dramatically since then—they were 4.0″ apart at the start of 2001—and continue to do so. Their position angle is rapidly changing, too, so the pair definitely looks a little different each year. They will reach a maximum separation of 7″ at the end of the twenty-first century, hit a secondary minimum of about 4″ in the twenty-third century, and won't return to their primary minimum separation until early in the twenty-fifth century. How small a telescope can you split Castor with on a good, steady night this year?

Both Castor A and Castor B are themselves tight, spectroscopic binaries (stars whose doubleness is detectable only by examining the chemical spectrum of their light). This is also true of Castor C, a star any telescope can reveal as a magnitude 8.8 orange point of light about 72" from Castor A and B.

As seen from midnorthern latitudes, Castor and Pollux are the brightest stars which form a truly close pair to the naked eye. As a matter of fact, the mere 4 1/2° separation between them means that they fit together in the field of most binoculars, or at any rate the lower, more standard power ones like 7x binoculars. No wonder then that this pair of stars so prominently placed in the zodiac became associated with the most famous twins of myth.

In Greek myth, they were called Castor and Polydeuces. Brothers of Helen of Troy, Castor and Polydeuces took part in many of the great adventures of their mythical generation (the generation of Hercules and many other illustrious heroes). Most famous of these adventures was the quest for the Golden Fleece. Castor and Polydeuces were Argonauts with Jason in that quest and it was on the return journey that a great storm raged around the ship Argo. Suddenly the storm calmed and above the heads of Castor and Polydeuces appeared two glows. Their fellow Argonauts took this as a sign that the twins had brought the blessing of the gods and saved the day. These glows, sometimes seen on the spars of ships, are what has become known as St. Elmo's Fire—a kind of slow, harmless electrical discharge which can occur after a thunderstorm has passed its peak. The tradition was established that the Twins should be invoked for safety at sea and the Romans adopted the custom, raising the twins they called Castor and Pollux to an even greater level of importance. The oath "O Gemini!" survived into modern English as "By jiminy!"

Gemini's Other Wonders

Gemini has two variable stars easily visible to the naked eye, a cluster seen through the edge of a cluster, and one of the most intense and dramatic of the planetary nebulas.

The two bright variable stars are Zeta Geminorum (Mekbuda) and Eta Geminorum (Propus). Mekbuda vies with Delta Cephei itself for the title of the brightest Cepheid whose variations are large enough to be easily noticeable. Mekbuda ranges from 3.6 to 4.2 over a period of 10.15 days. You can tell it is near minimum when it is markedly dimmer than Delta Geminorum (Wasat), the magnitude 3.5 star a few degrees northeast of it. The other bright Gemini variable is Propus,

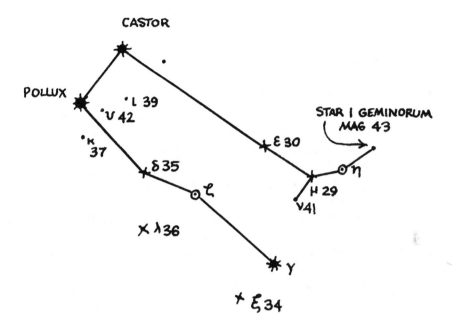

Figure 5.2. Map of Gemini identifying the naked-eye variable stars Eta and Zeta Geminorum (indicated by dots with circles around them) and the magnitudes of unvarying stars with which to compare them (decimal points have been removed from the magnitudes so "37," for instance, means magnitude 3.7).

located in the northern feet of the Twins. In contrast to Mekbuda, Propus is a semiregular long-period variable, its brightness ranging from about 3.2 to 3.9 in a period of 233 days. That is from almost as bright as its neighboring star Mu Geminorum (Tejat) to almost as dim as its neighboring star Nu Geminorum.

Just north of the foot of the northern Twin formed by Tejat, Propus, and 1 Geminorum is the giant, bright open cluster M35. If you have a reasonably dark sky, you can see this high and conveniently placed cluster plainly as a fuzzy spot with the naked eye. M35 is about 1/2° wide and glows at about magnitude 5.1. It is near the ecliptic, so is sometimes hidden by the Moon, and often passed closely by planets. This big cluster needs low magnification in telescopes to look its best, and that is quite splendid indeed. If you look closely with a 6-inch telescope and slightly higher magnification in dark skies you should also see in the field of view a hazy little patch of light. This is NGC 2158. It is only 5' across and 8th magnitude and lies on or just beyond the outer fringes of the southwest part of M35. You should ideally have a telescope larger than a 6-inch if you want to start seeing individual 13th-magnitude stars in NGC 2158. The fascinating thing is that NGC 2158 is really about the same size as M35— it just happens to lie about six times farther from us. M35 is about 2,800 light-years from Earth, NGC 2158 about 16,000. Also interesting is the fact that when we look at these clusters we are looking just a few degrees from the galactic anticenter—that is, the point in the heavens in the exact opposite direction of the center of our galaxy.

Not far southeast of Delta Geminorum (Wasat) burns NGC 2392, the Eskimo Nebula. This planetary nebula gets its name from the fact that on long-exposure photographs its shape somewhat resembles that of a face with the fringe of a parka hood around it. The nebula looks even more complex and strange, like some kind of bizarre mask, in Hubble Space Telescope images. But even small telescopes show the nebula plainly and through a large amateur telescope its features burn so bright that it looks like some kind of special effect from a science-fiction movie.

FEBRUARY'S OBSERVING EXPERIENCE

By day in February, there are the first encouraging signs that spring may not be infinitely far away. You might see crocuses pushing up from the hard soil. You might hear some increase in the amount of birdsong from a few species. And if you watch astronomical phenomena, you notice that the rate at which the day is lengthening begins to accelerate (the time of sunset is getting rapidly later) and the Sun is taking a higher path across the sky than it did in December and January.

By night in February, it is too soon for most of us to hear the first frog choruses of spring or much else in the way of nocturnal music from wildlife. So our nighttime environment may continue cold and silent, with little more than the crunch of ice or snow under our feet.

If you observe very far from city lights, a February experience that you can strive for starting about 90 minutes after sunset is the zodiacal light, a tilted, tapering tower of faint light that in February extends most steeply as much as 30° or even 50° to the upper left—along the zodiac—from the last of the Sun's afterglow. A similar sight of the zodiacal light at dawn's first gleaming can be best seen (because once more it is steepest) in October. The cause of the zodiacal light is sunlight reflected from meteoroidal dust throughout the inner solar system but especially in the plane of Earth's orbit.

As I noted at the start of this chapter, February is the snowiest month of the year for most of us. In addition to obvious problems like slipping in the snow, getting wet trudging through it, and having trouble getting a telescope out of a shed that has snow drifted up against the doors, there is the problem of light reflected from snow cover hampering your dark adaptation. Don't think this latter problem would be significant? It could wipe out all faint stars from your naked-eye view, causing you to lose as much as a couple of magnitudes from your naked-eye limit.

Despite the problems which snow creates, however, it can also be a force for creating beauty in your winter skywatching experience. Far from city lights, you can sink soul-deep into seeing Orion and Sirius hanging between the softly molded shapes of snow-hooded evergreen trees. You may also have nights when part of the sky is bright with stars while snowflakes are still drifting down around you. I remember one particular evening when there was a single, weak incandescent streetlight well behind me. I was looking straight up at a night sky clear to the sixth magnitude when suddenly I witnessed an impossible marvel: it seemed that stars were drifting down out of Auriga toward my upturned face. My moment of astonishment was replaced by silent wonderment at the snowflakes falling from among the stars.

Chapter 6

March

[Constellations covered: Leo; Cancer; Leo Minor, Lynx, Sextans; Hydra (front section); Carina, Puppis, Vela]

For most of the world's population, no month brings such changeable weather as March. As astronomers know, Earth in March is reaching the vernal equinox part of its orbit, where the planet's axis is at right angles to the incoming rays of sunlight. Thus all the world is lit similarly, with days and nights of equal length. However, when days get rapidly longer and the Sun rapidly higher at midnorthern latitudes, the input of solar energy causes strong temperature differences and helps create powerful storms and winds. In the U.S., the greatest blizzards have occurred in March. Yet often just a few dozen miles or a few hours separate mighty snowstorms from outbreaks of severe thunderstorms and tornadoes.

For an astronomer it is hard to know from night to night in March if you will be wearing a parka or a light jacket. What you can be sure of, though, is the prominence of bright and noble Leo the Lion, the Beehive Star Cluster in Cancer, and the head and heart of horrible Hydra the water monster. You can count on seeing from midnorthern latitudes the highest "young moons" and often having your best chance of the year to see the bright but elusive planet Mercury. You also have a better-than-usual chance to behold the sky flame with a great Northern Lights display.

Our March all-sky map (in the back pocket of this book) shows that in the middle of March evenings the bright constellations of winter (and constellations of late fall in the northwest sky) have moved on and are now concentrated into a kind of vast bright crescent, the western third of the sky. Neither Sirius, Orion, Taurus, or Perseus are setting quite yet, but the last part of Andromeda is setting and Cassiopeia is getting low in the northwest. On the other hand, a north circumpolar pattern even more brilliant than Cassiopeia—the Big Dipper—is now very high in the northeast, with the nose of the Great Bear (Ursa Major) already touching the sky's central meridian. The curve of the Big Dipper's handle points to a great stellar sign of spring: the brilliant star Arcturus, now plainly visible, though still rather low, a bit north of east. Lower in the southeast, just becoming visible (if your treeline or city skyline is not too high) are the second brightest star of the spring constellations, Spica, and the modestly bright but compact form of Corvus the Crow.

But all of those sights are ones which are best displayed in other months. Our current chapter focuses on our featured constellations of March: Cancer; the front section of Hydra; Puppis and Vela; Lynx and other faint patterns; and, first and foremost, Leo the Lion.

LEO

Leo can represent any of the great lions of legend. He is most often associated, however, with the rampaging Nemean Lion that Hercules killed as the first of his great Labors. The lion's skin became the hero's cloak, as can be seen in depictions of the constellation Hercules.

The constellation Leo consists of two major parts. The back half, or hindquarters, is a fairly prominent right triangle of stars. (We will discuss it in a moment.) More famous and prominent is the Lion's front half, which is formed by an asterism that looks like a backward question mark and is often called "the Sickle." The handle of the Sickle (or dot of the backward question mark) is the first-magnitude star Regulus. But in the picturing of the Lion's form, the hook of the Sickle is the head, curving outline of mane and chest of the beast— and Regulus is the lion's heart.

Regulus and Other Leo Stars

Of the brightest stars, Regulus ranks twenty-one and is the dimmest of the 1st-magnitude stars. Yet Regulus is actually more famous and observed than some of the stars that are brighter than it. That is not just because it marks the bright southern end of a prominent asterism and heart of one of the brightest constellations, which really does look quite a bit like the outline of the creature it is supposed to represent. It is not even just because in the lore of ancient times the brightest star in "the king of beasts" would naturally be considered a royal star associated with the births and fates of kings. No, Regulus's greatest claim to fame is that it is the 1st-magnitude star closest to the ecliptic—the plane of Earth's orbit—and therefore the most frequent participant in spectacular close conjunctions with the Sun, Moon, and planets. Regulus is one of only four 1st-magnitude stars that can be occulted (hidden) by the Moon in our era of time (the others are Spica, Antares, and Aldebaran—not many thousands of years ago precession made it possible for Pollux also to be occulted). The only occultation of a first-magnitude star by a planet that has ever been observed (such an event requires a telescope) was that of Regulus by Venus on July 7, 1959 (the next such event will not occur for a few hundred years). When a slightly yellow-white planet like Venus or Jupiter comes close to slightly blue-white Regulus, the colors of both objects are enhanced by contrast.

Incidentally, the name Regulus itself means "little king." Since royalty are often called "blue bloods," it seems appropriate that the heart of the king of beasts is marked by a blue-white star.

Regulus is a B7 star located about 78 light-years from us. Its luminosity is as great as about 140 Suns. Interestingly, all the stars which make up the main pattern of Leo are fairly close (within a few hundred light-years of Earth) except the modest-looking one just above Regulus in the Sickle. This star, Eta Leonis (magnitude 3.5), is actually thought to be about 2,000 light-years away and may have a luminosity something like one hundred times as great as Regulus.

Other than Regulus, the most important star in the Sickle for skygazers is Leo's second-brightest star, magnitude 2.0 Gamma Leonis, which is also known as Algieba. The famous Leonid meteors of November have their radiant (their apparent point of origin) quite close to Algieba when they are at peak. But what is important about

Algieba itself is that it is a spectacular, though rather tight, double. The magnitudes of the components are 2.3 and 3.5. The spectral classes are not much different—the brighter star is K0, and fainter is G7. (Recall that the higher the number of a spectral class, the cooler it is within its spectral type—so, since K is the next cooler spectral type after G, a G7 star is only three classes hotter than a K0.) Many observers do indeed see both components of Algieba as a similar gold or gold-orange. Some find a hint of green in the fainter (but slightly hotter) star. What do you think? To find out, you will have to split this pair, which were 4.4″ apart at the start of the twenty-first century and will only widen a tiny bit more during the entire century.

By the way, don't confuse Gamma Leonis (Algieba) with another famous double star of spring, Gamma Virginis (Porrima). The latter is also a tight yellow pair, but as you will read in the May chapter, is too tight a pair to split in many amateur telescopes in the first decade of the twenty-first century.

The right triangle which makes up the hindquarters of Leo is a full hour of right ascension east of the Sickle. The brightest star of that triangle is magnitude 2.1 Beta Leonis, also known as Denebola, which means "tail of." In ancient times the tail of Leo, or rather the tuft at tail's end, was originally a naked-eye star cluster. But that star cluster got appropriated to become a constellation in its own right, Coma Berenices—Berenice's Hair (see May chapter).

A final important sun in Leo is the long-period variable star R Leonis. This somewhat ruddy object is a huge red giant which varies from about 10th to 5th magnitude (even as bright as 4.4 sometimes) over a period of approximately 312 days. R Leonis is about 5° west of Regulus, very near the dim naked-eye stars 18 and 19 Leonis. You'll need a good star atlas to locate it most of the time, when it is faint, but you can check the monthly listing of variable star maxima in *Sky & Telescope* to find out when it is next predicted to reach its peak brightness.

Galaxies of Leo

Got a telescope? Then Leo has some lovely galaxies for you. The most famous of these are two groups of three galaxies each. I call them the Leo Trios. One trio is located about 8° due west of the other. Both trios belong to a larger arrangement in space, the Leo Galaxy Group.

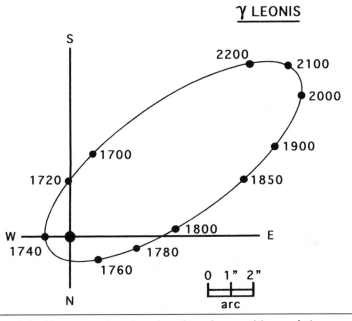

Figure 6.1. Diagrams showing the changing positions of the companion stars in the Gamma Leonis and Gamma Virginis systems. Note that Gamma Leonis B (the companion) is changing its position little in the twenty-first century and is at the far end of the orbit from its primary but Gamma Virginis B is changing position rapidly and at the near end of its orbit in the early twenty-first century.

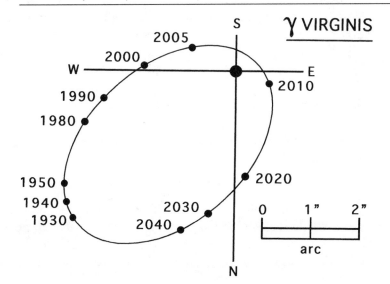

All six of these galaxies lie at roughly 26 million light-years from Earth. At this distance, the 8° separation between the two trios is more than 3 million light-years.

The first trio is located south of the belly of the Lion and is M95, M96, and M105. All its stars are between magnitude 9 and 10. The first is a spiral, the second a barred spiral (that is, it has a bar-shaped structure of stars across it), and the third, about 3/4° north of the others, is an elliptical galaxy. Can you detect a fainter galaxy, NGC 3384, only 8' east of M105?

The second trio is only a few degrees southeast of the star Theta Leonis, which is also variously called Chort, Chertan, and Coxa. The last of these names means "hip." This star is the more southerly of the two leading western stars of the Leo hindquarters right triangle. The trio of galaxies near it are M65, M66, and NGC 3628. They all shine at between 9.0 and 9.5. All are spiral. Two of them are considerably tilted toward us but the third, NGC 3628, is seen almost edgewise.

Another fine galaxy to look for in Leo is much less famous, though it may actually be a bit brighter than those of the Leo Trios. It is big NGC 2903, located several degrees southwest of the nose of Leo.

CANCER

Cancer the Crab has been described as the blank area of sky between Gemini and Leo. It is indeed one of the dimmest zodiac constellations. Nevertheless, in dark country skies some of its stars are easy to spot. The brightest of them is magnitude 3 1/2. And at the center of Cancer is a marvelous star cluster that is half a magnitude brighter.

The Beehive Star Cluster

The huge star cluster in the center of Cancer is M44. Amateur astronomers often call it by a nickname—the Beehive.

The view in a small telescope at low power shows why. M44 contains numerous stars of similar brightnesses (the bees) in bunches and pairs congregated in a large roughly elliptical area (the hive). The cluster is well over a degree wide and sits just north of the ecliptic, so it is a prime target not just for the Moon but for planets, which can spend one or more days passing through it. Eleven stars in the Beehive

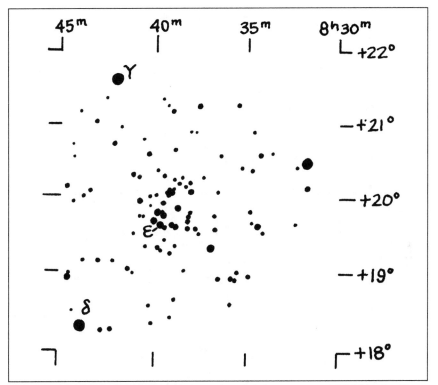

Figure 6.2. Epsilon Cancri and the other brightest stars of M44, the Beehive Star Cluster, are gathered in the center of this chart, which also shows the surrounding few degrees of sky. (Adapted from a diagram by Guy Ottewell.)

glow at between 6.3 and 6.9, with about seventy more brighter than 10th magnitude. A few of the brightest Beehive members have been just barely glimpsed as individual points of light amid the combined glow of their neighbors by sharp-eyed observers under the finest of sky conditions. Yet in merely very good skies, most of us can at least notice a fascinating unevennness of the glow across the face of the cluster, as if individual stars were on the verge of resolution. Any pair of binoculars will reveal the true nature of the naked-eye fuzzy patch of glow.

How easy is M44 to see with the naked eye? Pretty easy in a clear, moonless sky unless you have quite a bit of light pollution—which so many of us do these days. In ancient times, skywatchers noticed that a little bit of haze or thin cirrus cloud could render M44 undetectable while nearby naked-eye stars still were visible. That's understandable considering that M44 is an extended object, its magnitude 3.1 total brightness spread over a patch of sky rather than being concentrated into a point. The ancients said that M44 being invisible when surrounding stars were still seen presaged the coming of a storm. They were sometimes right because storm systems are often preceded by large areas of light cirrus clouds, which would tend to hide an extended glow while still allowing sight of pointlike stars.

When you look up with the naked eye or binoculars you find the Beehive just to the left (east) of being centered directly between two stars of Cancer—Gamma and Delta Cancri. These are, respectively, Asellus Borealis and Asellus Australis—the Northern Ass and the Southern Ass. There is a story behind this. Ancient people pictured the two stars as asses stepping up to a manger (the cluster) to feed. The Greeks knew the Beehive, M44, as Praesepe—the Manger.

The reason M44 appears bright and big is its relative closeness. Recent estimates place it at only 580 light-years away—closer to us than all but a few clusters. Don't forget that to fit the whole huge cluster into one telescopic field of view, let alone make it look concentrated and rich, will require a wide-field telescope at low power.

Other Wonders of Cancer

M44 is not the only impressive star cluster in Cancer. Four or five times farther away in space is a mighty cluster of great richness which also happens to be one of the few most ancient of all the open clusters. This is M67.

M67 is believed to contain as many as five hundred stars, a few times more than M44—and they are packed into an area only about 25' or 30' across. According to Steve O'Meara in his superb book *The Messier Objects*, the density of stars in the core of M67 is twenty-seven stars per cubic parsec (about 3.26 light-years). The Sun has no stars within a parsec of it. Of course, at M67's rather large distance, its individual stars shine at 10th magnitude (brightest members) to about 16th (dimmest). So it is not hard to see many of the brighter stars of M67, but some of the fainter ones appear as a background glow unless you are using a fairly large amateur telescope.

The total apparent magnitude of M67 is frequently listed as 6.9. But I have often enjoyed seeing it with the naked eye and am sure it is considerably brighter than that. Steve O'Meara estimates it at 6.1 and suggests trying to see it and M44 with the naked eye at the same time while thinking about their relative distances. M67 is about 9° almost directly south of M44 so the very widest-field binoculars can just fit them together in one view. M67 is very conveniently located for either naked eyes or optical aid—it is situated just 1.8° west of magnitude 4.3 star Alpha Cancri (also known as Acubens).

Many of our brightest, richest open clusters are only tens of millions or even just a few million years old. M67 is thought to be as much as 10 *billion* years old. It, like the globular clusters of our galaxy, has existed for most of the history of the universe. Our Sun and solar system may be less than half the age of M67. Most open clusters begin to drift apart in much shorter periods of time, due to gravitational tugs on the members as they pass near other stars in their orbit around the galactic center. Why hasn't this happened to M67? Because M67 has an orbit considerably inclined to the equatorial plane of the Milky Way Galaxy and therefore seldom is near enough to other stars (or clouds of gas and dust) to get disrupted. When we survey the heavens we notice that open clusters are plentiful near the Milky Way band (which roughly marks the galaxy's equatorial disk of stars) and cease when we get far enough away from that band in the sky. M67 is indeed quite a few degrees away from the Milky Way band.

But so is M44. Why is the Beehive, at an estimated age of only 400 million years, already becoming looser than M67? We need to remember that M44 is much closer to us than M67 is, so M44's angular separation from the Milky Way band (or, more precisely the galactic equator) works out to a much smaller distance in light-years above the galaxy's equator. At M67's distance from us of about 2,600 light-years, its angular separation from the galactic equator works out to its being a full 1,500 light-years above the galactic equator right now. That's far enough to avoid most of the "perturbing" effects of other stars.

So much for the two great and very different open clusters of Cancer. We can end our survey of Cancer by mentioning two of its fine double stars.

Actually, Zeta Cancri (also known as Tegmine) is a triple star. But the easy thing to do is to split it first into two components, magnitude 5.0 and 6.1, that are 6.2″ apart. The real trick is separating the first of these two stars into a magnitude 5.6 and 6.0 duo. The 5.6 and 6.0 pair were only 0.9″ apart in 2001, and though widening are doing so slowly enough that they won't until around 2020 reach a maximum separation of still only 1.2″. Splitting this AB pair therefore will require about a 6-inch telescope in the opening decades of the twenty-first century. This is one of the most rapidly orbiting of all visual doubles, however. The period of B around A (or rather around their center of gravity) is only sixty years, and at minimum separation the stars are only 0.6″ apart.

Remarkably, all three of Zeta Cancri's visible components are yellow and not much different in spectral class from the Sun. All three are also not greatly brighter than the Sun. Zeta Cancri A and B are roughly as far apart as the Sun and Uranus. Zeta Cancri C is much farther out in the system, taking about 1,150 years to revolve around the A-B pair.

If you have trouble splitting Zeta Cancri A and B, a much easier target is Iota Cancri. James Mullaney has called it "the Albireo of Spring" because of its resemblance to the famous wide gold and blue double of summer.[3] Even very low magnification reveals the two components of Iota Cancri, for they are 30″ apart. They shine at 4.2 and 6.6 and have been called "pale orange and clear blue" and "crocus and violet."

Minor March Constellations: Leo Minor, Lynx, Sextans

Cancer may be dim but it has those bright, mighty star clusters and superb double stars. It is also an important constellation due to its zodiac status. Three other March constellations can only be said to be faint, with just a few deep-sky objects bright enough for novice stargazers among the three of them.

Leo Minor (the Little Lion) is a few stars and small area north of Leo and south of Ursa Major, the giant bear constellation of the Big Dipper we look at in the April chapter. Just to identify the main stars of such an obscure constellation and trace its pattern is a good accomplishment. (Indeed, I encourage the reader to identify even the faintest constellations—they each have a fascinating place in astronomical history and each is, after all, a piece of the real estate of the heavens, albeit entirely by human convention.)

Leo Minor was a seventeenth-century invention of the astronomer Johannes Hevelius. It is the only northern constellation which has no

Alpha star at all! There is a Beta Leo Minoris, but it is outshined by magnitude 3.8 star 46 Leo Minoris. 46 Leo Minoris actually has a proper name: it is called Praecipua, which means "principal." But being the principal star of Leo Minor is not so big a deal!

A more interesting and important constellation invented by Hevelius is Lynx, so named because Hevelius said that one had to have the eyes of a lynx to see any pattern at all among the faint stars of this constellation! Actually, Lynx fills a long space of sky north of Gemini, Auriga, and Cancer and has several interesting deep-sky objects. 12 Lyncis and 19 Lyncis are both fine triple stars, the latter an easier split with much great color contrast. 38 Lyncis is a lovely double star. The most intriguing deep-sky object in Lynx is NGC 2419—a globular star cluster known as the Intergalactic Wanderer. This cluster is only about magnitude 10.4 and 4' across but it is located 300,000 light-years away. If it is actually under the control of our Milky Way galaxy it is the farthest out of the galaxy's globular clusters which can be detected by amateur telescopes. A final notable deep-sky object in Lynx is NGC 2683. It is a nearly edge-on spiral galaxy. It is magnitude 9.7 and measures 9' by 2'. For a detailed view of it, at least an 8-inch telescope is desirable.

An even better edge-on galaxy shines in the final of our three dim March constellations. Sextans the Sextant was also created by Hevelius. Its magnitude 9.2 and 8' by 3' NGC 3115 has been called the Spindle Galaxy. NGC 3115 is intense enough to look good in surprisingly small telescopes and an 8-inch at high power can show a starlike nucleus in it.

The brightest star in Sextans is magnitude 4.5 Alpha Sextantis. Sextans is due south of the Sickle of Leo, and covers an area not much larger than the Sickle. It is also due east of the fairly prominent head and frontmost body section of our next major constellation, Hydra. By the way, a good way to locate the Spindle Galaxy is to take the second-magnitude heart of Hydra, Alphard, and go several degrees almost due east until you find Gamma Sextantis—only magnitude 5.1 but much brighter than any star within a few degrees. If you go a few degrees farther east on the line from Alphard through Gamma Sextantis, you will come to NGC 3115.

HYDRA

Hydra is the Sea Serpent or Water Serpent. In mythology it was a specific kind of mythological creature which Hercules fought as the second of his Twelve Labors. Called the Lernean Hydra, it was a serpent with many heads and the number of them kept increasing. For each time a warrior cut off a head he would find that two heads grew back to take the original's place. And one of the hydra's heads was simply immortal, there was no way to kill that one. The solution of Hercules was to cut off each mortal head and have his nephew Iolaus immediately burn the stump with a torch—a cauterizing which prevented any heads from sprouting back. The immortal head could not be subdued even with fire. So Hercules rolled a giant boulder on top of it, rendering it helpless forever.

In the heavens, Hydra has only one head but is appropriately the longest of all constellations in the east-west direction. At the time of our all-sky map for the middle of March evenings, the head of Hydra has just passed the sky's central meridian and is therefore almost due south. But this monster's body stretches all the way to the southeast horizon and there is still more of it to rise. At midnorthern latitudes it takes well over seven hours for the entirety of Hydra to rise. When the tail finally comes above the horizon, we find that Hydra extends about two-thirds of the way across the southern sky. Although Hydra is a skinny constellation, modern constellation boundaries make it the largest in area of all the constellations.

Having noted all this about the length and size of Hydra, I must hasten to add that it is a surprisingly dim constellation on the whole. And yet the front section of the monster, which we study this month (the rest of Hydra we cover in the April and May chapters), features a bright heart and a modestly bright but compact and very noticeable head.

Indeed, the star which marks the heart of Hydra is a very prominent and memorable one. It is Alphard, which we noted above as a guide to the Sextans galaxy NGC 3115. Alphard is a distinctly orange star (through binoculars or telescope, at least). It also shines at a magnitude 2. That would be pretty impressive anywhere in the heavens, but it is especially so in the vast dim section of sky in which Alphard is located. Appropriately, the name Alphard is Arabic for "the solitary one."

Of course, there are some bright stars not all that far to the north-east and northwest of Alphard. It is more like a last outlier of these bright regions, ruling over a seeming near-emptiness to its south, southeast, and east. The brilliant stars which can be linked with Alphard are Procyon and Regulus. Those two form a large nearly per-fect right triangle with Alphard.

Alphard is the heart of the lengthy hydra. But Procyon can also be a guide for finding the monster's head, for the head lies less than 10° due east of Procyon. The head of Hydra does indeed look like the compact one of a snake. And although the head's stars are of only third and fourth magnitude, the concentration of them into a small area makes them conspicuous.

In the middle of March evenings, the head of Hydra and Procyon are the upper left and upper right points respectively of an equilateral triangle whose downward-pointing apex is the big magnitude 5.8 open cluser M48. M48 was one of the "lost" Messier objects until modern researchers decided that it must be this beautiful cluster, 4° exactly south of the empty position which Messier must have recorded mistakenly. M48 lies just within Hydra, near the constella-tion's boundary with Monoceros.

A marvelous object in Hydra which was little known to most ama-teurs before the past decade or two is the planetary nebula NGC 3242—"the Ghost of Jupiter." Its overall magnitude is about 8.6 with an 11.4-magnitude central star. In small telescopes, it shows a pale bluish disk roughly as big as Jupiter. With apertures of eight inches and larger, its amazing internal structure becomes apparent, including a ring which makes the nebula begin to look like an eye. The Ghost of Jupiter is located just 1.8° south of Mu Hydrae, the eastern member of a triangle of three stars which lie directly south (but far south) of the Sickle of Leo. East of Mu Hydrae is a brighter star, Nu Hydrae. They form the base of an upright equilateral triangle at whose top shines one of the reddest stars in the heavens, U Hydrae. It varies semiregularly over a period of about 450 days from about 4.8 to 6.5, so binoculars or a small telescope will be needed to appreciate this cool carbon star's color.

CARINA, PUPPIS, VELA

We noted in the February chapter that the Gemini twins were among the many heroes who set sail with Jason on the Argo, the greatest ship of Greek mythology. The ancient Greeks assembled the hugest of all constellations to form the constellation Argo. By the nineteenth century, however, the telescope had revealed so many stars and deep-sky objects within the bounds of Argo that it was becoming unwieldy. Astronomers started using three major component parts of Argo that had been invented in modern times: Carina the Keel, Puppis the Poop(deck), and Vela the Sails. These three parts of Argo replaced the ship officially in 1930 when the International Astronomical Union voted to formalize those constellation boundaries completed by Eugene Delporte. Also made official, just east of Puppis and north of Vela, was the faint constellation Pyxis the (Box) Compass (supposedly of Argo)—which had been made from the stars of the abandoned Malus the Mast (also supposedly of Argo).

Canopus and Carina

In the February chapter, we discussed the foreward light of the ship, the second-brightest of all stars, Canopus. Canopus was once Alpha Arguis (the Alpha of Argo) but is now Alpha Carinae. For it lies in Carina the Keel, one of the brightest and richest of all constellations in both stars and deep-sky objects. Carina is the dwelling-place of many wonders. It has magnitude 1.7 Miaplacidus (Beta Carinae) and magnitude 1.9 Avior (Epsilon Carinae). It has the magnificent Eta Carinae Nebula, somewhat less intense than the Great Orion Nebula but bigger and by southern observers favored over the latter. It has Eta Carinae, a now-faint star which is distant, superluminous, probably primed to go supernova, and which at one point in the nineteenth century brightened to slightly outshine even Canopus! Carina also has IC 2602, a star cluster which only falls moderately short of deserving its nickname: the Southern Pleiades. Carina, however, lies entirely below the horizon for viewers at 40° N latitude. So I will limit my commentary on it here to these mere mentions of its most stunning wonders.

Puppis

The other parts of the old Argo are more accessible to northerners. Part of Puppis forms the entire east boundary of Canis Major, and part of it extends to the south and southeast from Canis Major. Even its southern boundary at least gets up to the horizon of observers at 40° N latitude. Vela is mostly due east of the bottom half of Puppis (though it does extend a bit farther south than Puppis), but many of its interesting sights can be seen, though admittedly low and dimmed, from most of the United States.

Puppis was mentioned in our February chapter where I advised referring to this chapter to read up on it and do some observing of it in February. In the northwest corner of Puppis are M46 and M47. They are about 15° east of Sirius and glow only about 1 1/2° apart. They are both open clusters about a 1/2° across and both among the sky's finest clusters. And yet they are fascinatingly dissimilar. The western cluster, M47, is rather messy, and shines with many bright stars (magnitudes 5 to 9). M46 is much more orderly and has more numerous though dim stars (magnitudes 10 to 13). M47 is supposed to have a total magnitude of about 4.4, M46 of 6.1. M46 is, however, the greater cluster in true size and brightness, for it lies 5,300 light-years from us, compared to about 1,550 for M47. Just north of the center of M46 is an amazing sight for medium to large amateur telescopes: the eleventh-magnitude, 1'-wide planetary nebula NGC 2438. This nebula is believed to be another 1,200 light-years farther from us than M46. We thus see it shining through the cluster from well behind M46.

M93 is a Puppis open cluster about 7 to 10° northeast of the Southern Canis Triangle (Delta, Eta, and Epsilon Canis Majoris). M93 lies only 1 1/2° northwest of third-magnitude Xi Puppis. It is about as bright as M46 with fewer stars but in a smaller space. The brightness of the individual stars of this lovely cluster are from magnitude 8 to 13.

Some observers think that the finest of all the Puppis clusters is NGC 2477. It was probably missed by Messier (and by many later listers of prominent clusters) because of its relatively southerly position (it is at –38°33'). Though only a respectable 5.8 in total magnitude, the cluster is an incredibly rich congregation of eleventh-magnitude and dimmer stars. About three hundred stars teem within its 27' diameter. NGC 2477 is also easy to find, for it lies only 2 1/2° northwest of bright Zeta Puppis.

Zeta Puppis is also known as Naos, and is a remarkable star. Shining at a magnitude of 2.2, Naos is about twenty thousand times more luminous than the Sun. It is also one of the hottest stars, and about as far to the blue end of the spectrum as stars get.

Puppis contains a very unusual variable star called L^2 Puppis. This semiregular variable has an average period of about 140 days and ranges from sixth magnitude at minimum to occasionally brighter than third magnitude. Unfortunately for many observers, this star is quite southerly (almost –45°).

Vela

The exciting objects in Vela do not get far above the horizon for viewers at 40° N latitude. Much better views are visible to observers at 30° or 25° N (southernmost U.S.), however. So at least two Vela sights are worth mentioning here.

The brightest star of Vela, and third brightest even in the vast former expanse of Argo, is Gamma Velorum. What a shame for northerners that its declination is –47°. First of all, the star splits even in binoculars into a spectacular white (or "white and greenish white") pair of magnitude 1.8 and 4.3 gems, 41″ apart. There are also lesser companion stars, of eighth and ninth magnitude, about 1′ and 1 1/2′ away from the bright pair. But it is the nature of the brightest member of the system which is most fascinating. This is by far the brightest and closest of the Wolf-Rayet stars, a class of extremely hot stars. It is also a variable star whose very slight variations in brightness occur in a period of 154 seconds!

Gamma Velorum was dubbed by medieval Arabic observers with a few of the names which unfortunately were also applied to a number of other bright, southerly stars (Suhail, Muhliphein, etc.). The name Alsuhail has gotten itself attached fairly firmly to Lambda Velorum, leaving brighter Gamma Velorum with no widely accepted proper name—until a few decades ago. That was when three names were slipped into usage to try to honor the three astronauts who died in the Apollo 1 fire. The astronauts were Virgil "Gus" Ivan Grissom, Edward H. White II, and Roger B. Chaffee. The order of letters in the names Ivan, Second (II in White II), and Roger were reversed to come up with

Navi, Dnoces, and Regor (this is the same trick played in the nine-teenth century to introduce names for two stars in Delphinus—see September chapter). Navi and Dnoces were applied, respectively, to Epsilon Cassiopeiae and Iota Ursa Majoris—two stars which already had proper names (Segin and Talitha Borealis). Regor was applied to Gamma Velorum—and seems to have become a fairly common usage.

As far north as possible in Vela, virtually right on its border with the very faint constellation Antlia the Air Pump, is one of the finest of the planetary nebulas: NGC 3132, the Eight-Burst Nebula. Only about 10° above the horizon for viewers at 40° N, its overall magnitude is 8.2 and that of its central star is about 10.

STEEP ELONGATIONS, NORTHERN LIGHTS, MESSIER MARATHONS

The months surrounding the time of the vernal equinox are the ones during which the ecliptic makes its steepest angle with respect to the western horizon around dusk. This means that an evening elongation (angular separation) of Venus or Mercury from the Sun is at this time of year as favorable as it can be. If Venus is, say, 40° from the Sun or 20° from Mercury this month, the line between them and the Sun is as vertical as it can be and thus places them as high in the sky as possible. For Mercury this is just about critical. If Mercury has a greatest elongation from the Sun in the evening sky in February, March, or April it is high enough to be more easily visible after sunset than at any other time of year for viewers at midnorthern latitudes. (Note: for viewers at 30° or 40° S latitude, Mercury is highest [at greatest evening elongation] within a few months of September, for that is the month of the Southern Hemisphere's spring equinox.) At its best spring elongations, Mercury doesn't set until almost two hours after the Sun as seen from midnorthern latitudes.

The same favorable effect of the steep ecliptic at nightfall is beau-tifully seen with spring displays of "young" crescent Moons. The "age" of the Moon is the amount of time that has passed since its invisible New Moon phase, a phase at which it is very near to the Sun in our sky. To get a one- or two-day-old Moon as high as possible as dusk is fading and the sky darkening around it, you need the steep posi-tioning that a lunar crescent has around early spring.

Around the equinoxes—both spring and autumn—is also approximately the time when you are most likely to see the Northern Lights, technically known as the aurora. In this case, the most active latitudes of the Sun happen to be pointed most directly at Earth in late winter/early spring and late summer/early fall.

Of course, even around these best times of year, the aurora is not often seen prominently to 40° N and farther south in the United States unless we are near the peak in the roughly eleven-year cycle of solar activity (or a few years after that peak). The last peak or solar maximum occurred in May 2000. Thus auroral activity should not be as extensive and south-reaching in the Northern Hemipshere in 2003, 2004, or 2005. Still, you can never tell when such a display—with its potentially colorful and moving patches, arcs, rays and curtains— might occur. The equinoxes are always a time to be more vigilant for outbreaks of aurora (an excellent online aid to knowing when to look for an aurora is www.spacewether.com).

March is not just the best time to see crescent Moons and inner planets high, it is one of the best times to see the aurora. It is also the only time you could conceivably succeed in seeing all those deep-sky objects of the famous Messier catalog in one night.

Every March, there are amateur astronomers who participate in such attempts to see all the Messier objects—or at least as many as they can. These events are called "Messier marathons." March is the only time when the Sun shines in the largest gap of southern heavens which does not contain a single Messier object—the gap between M30 in eastern Capricornus (farther east and lower in the sky than M72 and M73 in Aquarius) and M74 in eastern Pisces. (M30 and M74 are highest at mid-evening in September and October, respectively, so refer to those chapters to learn more about these objects.)

On March nights of Messier marathoning, M74 is the first object you must look for, low in the dusk, and M30 the last, low in the dawn. M74 is difficult enough to see sometimes even when it is high in a fully dark sky as it is in October. Thus, many Messier marathons begin with missing M74 and the knowledge that the full list will not be achieved. That failure (if it occurs) is one that observers must accept so that they can move on to enjoy plenty of other wonderful and rewarding sights while honing their skills in the rest of the night, for that is what these marathons are really all about, or should be.

THE MARCH OBSERVING EXPERIENCE

March is the most capricious of the months and can even be the most virulent—that is to say, objectionably harsh and strong to observers. It is the month of the greatest blizzards, of some of the worst tornado outbreaks (especially in the American South), even of the peak of flu outbreaks. One March I experienced a terrible flu, a tornado watch, a great auroral outbreak, a violent sleet which fell clattering to cover the ground to a depth of four inches, and some of the pleasantest days of gentle sunshine and breezes imaginable—all within a week or so! All of these things except the last were violent or wild, and several of them were also eerily beautiful. It seems that the year of the natural world dies gently and quietly in autumn (though dramatically—with colored leaves and bird migrations). But it is born violently and tumultuously in spring.

Of course, starry March nights are usually more peaceful than any of the above phenomena of spring. But there are still enough stimulants of these nights, relatively subtle though they may be, to stir us inside with the expectation of spring. For instance, most of us will experience in March a first night when no gloves or hat, and only a light jacket, are needed if we are not going to be out for a really long time. The warmer day preceding such a night has opened up smells of the soil and other parts of nature which now become part of our stargazing for the first time in months. There may be a number of flowers blooming in our yard or wherever else we are. March is a windy month, but at least we can relish the lack of any truly severe chill in its wind. Red-wing blackbirds have become active in wetland areas by day. Grackles and robins return. But you may not hear many birds at night at this time. Instead, there may well be in the country on your warm March nights the music of creatures who sang hundreds of millions of years before the first birds: the frogs may add their choruses to the evening. Where I live, the tiny frogs called spring peepers will be making themselves heard.

What's most remarkable is that if you are an astronomer, the arrangements of the stars may themselves bring to mind all the senses and feelings of spring, even when these other sensations are not really present to your ears, nose, or skin. When early in the evening the Beehive star cluster hangs high, and Leo is getting lofty, and the compact and handsome little head of Hydra is seen well above the long, long monster's lonely orange heart, when these and other celestial sights are seen, then the conviction that spring is coming—or has arrived!— is felt, lifting the spirits in your heart.

Chapter 7

April

[Constellations covered: Ursa Major (including the Big
Dipper); Canes Venatici; Corvus, Crater, Hydra
(middle section), Crux (briefly)]

W hen we look high in the south in the middle of April
evenings, we still see Leo the Lion. His head, mane, and
chest have left the central meridian, but his hour-later hindquarters
are now about to pass that important line. But this is a time when we
are very likely to turn north under a starry sky. For high in that direction is another fierce and huge beast, represented by another bright
and even huger constellation. The beast is Ursa Major the Great Bear.
Yet beginners mainly focus on only a part of this lumbering denizen
of the north. Part of the Great Bear is also an asterism—not by itself
an official constellation—which is probably the most famous of all by
name in modern times. This is true even though the name is different
in countries as closely related as the United States and Great Britain.
The asterism is known in Britain as "the Plough." It is known in the
United States as "the Big Dipper."

The Big Dipper is theoretically visible at any time of the night
during any time of the year if you live at 40° N or farther north. In
practice, however, there are seasons and times of night when the
mighty pattern is low in the north, lost behind trees or buildings,
dimmed by horizon haze or light pollution. April is the month when
the Big Dipper is at its best. The striking pattern is in April at its highest
in the north in the convenient hours right after nightfall. Interestingly,

when it is highest is exactly when the imagined shape of this vessel for dipping and pouring appears upside-down. You might imagine that the Big Dipper is spilling its contents and producing April showers.

What does the rest of the starry sky look like when the Big Dipper is at its peak altitude? Our all-sky map shows that Sirius, Orion, Taurus, the Pleiades, and Perseus are finally setting or about to set. In fact, for viewers around 40° N, all of these will be lost or nearly lost in bright evening twilight before April is over. Nevertheless, a higher layer of departing winter constellations—Auriga, Gemini, and Canis Minor with Procyon—is still easily visible in the west. Meanwhile, in the east and southeast, the two brightest stars of the spring constellations, Arcturus and Spica, are in clear sight, already part way up the sky. As a matter of fact, very low in the northeast, Vega, the most brilliant star of summer, has just risen into view.

We already note Leo high in the south and Ursa Major high in the north. But dimmer constellations near these are among our featured subjects this month. Right under the extended handle of the Big Dipper is Canes Venatici the Hunting Dogs and not far to the right of Spica (and far lower left of Leo) is Corvus the Crow, Crater the Cup, and a middle segment of long Hydra. Canes Venatici harbors a few of the most marvelous galaxies in the heavens. It shines just above the constellations with the most galaxies visible to amateurs—Virgo and Coma Berenices. You can observe Virgo and Coma Berenices and their great gatherings of galaxies with a telescope on April evenings, but for organizational purposes we will be discussing them in our May chapter—where we also examine Spica, Arcturus, and the stars of Virgo and Boötes.

For now, however, get ready to set your clocks forward an hour, watch for the ancient Lyrid meteor shower, and study the following constellations which are at their best on April evenings: Canes Venatici, Corvus, Crater, the middle section of Hydra, and Ursa Major with its asterism the Big Dipper.

URSA MAJOR AND THE BIG DIPPER

The Big Dipper is more than just a bright star pattern. It is a compass, a clock, a yardstick, and a road sign. Before discussing these amazing functions, however, let us consider the stars of the Big Dipper in their own right.

Stars of the Big Dipper

Seven stars make up the Big Dipper. None is quite a 1st-magnitude star, but six of the seven are 2nd-magnitude. From the lip of the bowl, around the bowl's bottom and on outward to the end of the handle, the Greek-letter designations of the Big Dipper's stars are the first seven letters of the Greek alphabet: Alpha, Beta, Gamma, Delta, Epsilon, Zeta, Eta. The proper names of these stars are (in the same order): Dubhe, Merak, Phecda (or Phad), Megrez, Alioth, Mizar, and Alkaid (or Benetnasch).

This is not the order of their brightness. What would most people say who have looked at the Big Dipper casually if asked—by day—which of its stars is brightest? Most would respond that they didn't recall, that all of its stars seemed of fairly similar brightness. This assessment is true for six of the stars, because the brightest of the six is only a bit more than half a magnitude brighter than the dimmest of the six. Still, half a magnitude difference in brightness is easy to distinguish when you actually look at the stars carefully. Making sure that all the stars are high in the sky (when the Big Dipper is low, some of its stars are lower than others and will be more dimmed by atmospheric extinction), check their brightnesses. You'll find that they break into two subtly different groups in brightness. One group is composed of Dubhe and Alioth at 1.8, Alkaid at 1.9, and Mizar at 2.0. Another, dimmer group is Merak at 2.3 and Phecda at 2.4.

But what of the seventh star of the Big Dipper? This is Megrez, the one which connects bowl and handle. It shines at only 3.3, almost a full magnitude fainter than any of the other Big Dipper stars. As a matter of fact, there are five stars in other parts of Ursa Major which are as bright or brighter than Megrez. Still, this dimmest of the Dipper seven serves its purpose. Being the midmost of the Big Dipper stars, it is not easily ignored in the striking pattern.

Of all the Big Dipper seven, the star most interesting to observe is undoubtedly the one at the bend in the handle—Mizar. If you have fairly good eyesight and look carefully at Mizar on a clear and steady night, you will soon discover the first point of interest: Mizar has a remarkable little companion star shining right beside it! Can't see it with your unaided eyes? Then even the weakest pair of binoculars or field glasses will reveal it. This is the most famous and enjoyable naked-eye double star in all the heavens. The companion star shines at magnitude 4.0 at a distance of 11.8' from Mizar. It is called Alcor, but another name for it among the medieval Arabs meant "The Test." Seeing Alcor was considered a test of vision, but probably not a test of unusually sharp vision. Those were the days before eyeglasses. Normal vision should be enough to detect Alcor under good conditions. If you don't succeed, however, don't be too disappointed. Try in a sky that is darker or a night when star images are steadier. And in the meantime turn a small telescope on Mizar itself and get ready for a wondrous surprise: Mizar itself is a wide and spectacular telescopic double.

The separation between Mizar A and B is 14"—almost exactly fifty times less than the gulf between Mizar and Alcor, but still very easy to view (splittable in a good 50mm refractor at 25× or so under excellent conditions). What's more, whereas Mizar A burns at 2.3, Mizar B shines at 4.0—the same brightness as Alcor. Furthermore, all three of these stars have almost exactly the same blue-white color and similar spectral class.

As a matter of fact, the spectral class of all five of the bright middle stars of the Big Dipper is between A0 and A3. At the end of the handle, Alkaid is a hotter, B3 star. At the lip of the bowl, Dubhe is a much cooler K0 star. Can you tell with your naked eye that of all the Big Dipper stars only Dubhe has a distinctive color? What is that hue? With binoculars or a small telescope, you can ascertain that Dubhe is orange.

The Big Dipper Cluster

These striking differences of Dubhe from the other Big Dipper stars in color, and Dubhe and Alkaid from the other Big Dipper stars in spectral type, bring us to a fascinating question. Are the stars of the Big Dipper really traveling through space together—are they truly a group?

Complete beginners to astronomy often assume that stars near each other in the sky are also near each other in space. But novices soon learn that two stars side-by-side in the sky are usually at vastly different distances. Even if they are of about the same apparent brightness—like the stars of the Big Dipper—one is usually a far more luminous sun that happens to be much farther away from us.

So once you've learned a little about stars your guess would likely be that the Big Dipper stars are probably not related in space. Of course, if they were members of a star cluster like the Pleiades, they would be going through space together, loosely bound to one another by their gravitational pulls. But the Pleiades, one of the very closest clusters, is still a quite compact arrangement of stars, only about 1 1/2° across. The Big Dipper measures just over 25° from end to end. So surely the Big Dipper stars are not related to one another?

Well, the truth surprises. As far back as 1872 it was confirmed that the middle five stars of the Big Dipper share a similar motion through space. The end stars are going elsewhere. The middle five stars form, along with eleven or more other stars in Ursa Major and neighboring constellations, a very loose and sparse Ursa Major Cluster. It is made up of mostly stars in Ursa Major but would perhaps still be more vividly called the Big Dipper Cluster. About a dozen of the stars are bright enough to glimpse with the naked eye. They are spread over an area more than 23° wide. In space, this cluster measures about 18 by 30 light-years, and is centered about 75 or 80 light-years away from Earth. It is the closest of all star clusters.

But the story doesn't end there. Astronomers have found stars possibly related to this cluster much farther afield in the sky. It's possible that the 2nd-magnitude lucida (brightest star) of the early summer constellation Corona Borealis is a member of the Ursa Major Cluster. Indeed, almost all the way across the sky from the Big Dipper there are stars which seem to share a similar motion with the cluster. These could be members of a much vaster Ursa Major Moving Stream. Among the stars that might possibly be members of this Stream are Delta Leonis, Beta Aurigae, Beta Eridani—and Sirius! Our own solar system could be passing through the outskirts of this Stream.

In the late 1990s it became possible to judge the distances of stars even hundreds of light-years away much more accurately by applying the parallax method (see chapter 2) to precise positional measurements performed with the Hipparcos satellite. The distances from Earth of the middle five stars of the Big Dipper are all about 80 light-years, give or take a few light-years (and at this distance the accuracy of the Hipparcos figures is good to within about plus or minus 1 1/2 light-years). Alkaid is 101 light-years from Earth, Dubhe 124 light-years—so clearly they are chance additions to our current Big Dipper pattern. I say current because, as astronomy books are fond to point out, given another fifty thousand years of stellar motions, Alkaid and Dubhe will move in different directions and the figure of the dipper will be severely deformed. In our thousands of years of history, however, it is amazing to consider how the central stars of the Ursa Major

Cluster are not just at similar distances from Earth but are also arranged in a line in space that is almost perpendular to our line of sight.

The Many Uses of the Big Dipper

The Big Dipper is beautiful just to look at and, as we have just found, fascinating to learn about. For a long, long time, however, the Big Dipper has also been employed for what we could call functional purposes. Perhaps that doesn't sound exotic or fanciful. Actually, though, when the function of something can be to help find your way home through a forest at night, or tell how many hours will pass before dawn, or locate a number of other stars and constellations, it is not boring. The Big Dipper can be used for all of the above adventurous (though also sometimes very practical, even conceivably life-saving) purposes . . . and more.

The compass function of the Big Dipper is the best known. Surely every Boy Scout and Girl Scout should learn how to use the Big Dipper to find the North Star and therefore the direction of true north. This is something many people learn who never learn anything else about the constellations. (Or at least that used to be the case. I wonder if even this lesson is being taught to people much in these recent years of so much light pollution and such a disconnection of humans from most aspects of the natural world.)

The key to finding the North Star is to use the two Big Dipper stars on the far side of the bowl from the handle. These stars are Dubhe— the lip of the bowl—and Merak, the star which forms the bottom of the bowl on that same side. Together, Dubhe and Merak are known as the Pointers. To find the North Star, you must draw an imaginary line from Merak through Dubhe and extend it until it passes near a solitary bright star. Not a 1st-magnitude star, but one of magnitude 2, similar in brightness to the brighter stars of the Big Dipper. That star is Polaris, the North Star. And, since Polaris lies very close to the north celestial pole, when you are facing Polaris you are looking almost exactly due north.

It all sounds quite easy, using the Pointers to find the North Star. And it *is* easy, as long as two things are true. First, as long as the Big Dipper is not hidden by objects in the landscape or by thick haze or light pollution down low when it is near the north horizon (which it is, for instance, on autumn evenings). Second, as long as you recall which stars of the Big Dipper are Dubhe and Merak and remember to draw the line from Merak through Dubhe (not vice versa). This latter requirement might be a little more difficult than you think if you consider that sometimes the imagined dipper-shape of the Big Dipper will appear upside-down, sometimes right-side up, sometimes standing on its handle, other times standing on its bowl with the handle sticking up. On spring evenings, the Big Dipper is far above the North Star and the orientation of the dipper is upside-down—so the line from Merak through Dubhe points down. On fall evenings, the Big Dipper is below the North Star, the dipper appears upright and the line through the Pointers is extended up.

The changing positions of the Big Dipper with respect to the North Star bring us to the second fascinating function of the great asterism as a clock and/or calendar. I have just alluded to where the Big Dipper is found in the middle of spring and fall evenings (or, more precisely, where the Pointers are found around 10 P.M. DT in April and October, respectively). In July around 10 P.M. DT, the Pointers are to the left of the North Star, and in January around 9 P.M. ST, they are to the right of the North Star. But where is the Big Dipper during the other hours of the night at these different times of year? The rule is pretty simple. It takes almost twenty-four hours for the Big Dipper, moving counterclockwise, to make a complete circle (actually about four minutes less than twenty-four hours—the "four minutes earlier" every night as Earth orbits the Sun, so that even circumpolar stars have their seasonal progression—see chapter 2). Therefore if you look out your window on an April night and find the Pointers horizontal with Dubhe exactly to the right of Merak (that is, the Pointers are due left of the North Star), then you will know that the Big Dipper has progressed one-quarter of its full circle around the pole. That would require one-quarter of about twenty-four hours, or six hours past when the Pointers were directly above Polaris at 10 P.M. DT: the time is about 4 A.M. DT. So you might say that the line formed by Merak, Dubhe, and Polaris is like the hour hand of a clock in the sky. But this is a twenty-four-hour clock in which the hand moves counterclockwise. And the number 0 for midnight would be at the top of the clock in March, but progress counterclockwise around the clock during the months that follow! (That last statement may sound confusing—numbers which themselves slowly move around the dial!—but the calculations we mentioned a few sentences back are quite simple.)

Finally, what about the Big Dipper as yardstick and road sign? The really useful primitive tool for measuring angular distances is your hand at arm's length. But the Dipper itself has some very convenient dimensions to refer to, and can be used to determine the size of your particular fist-at-arm's-length more precisely. The distance between the two Pointer stars is about 5°; across the top of the bowl is about 10° (the average width of a person's hand held at arm's length); and in a straight line from one end of the Big Dipper to the other (from Dubhe to Alkaid) is about 25°. As for "road sign" . . . well, I mean indicator of the direction to various other constellations and stars. The line through Merak and Dubhe points the way to Polaris. Drawn the other way it points to the middle of Leo. The curve of the Big Dipper's handle extended (about one Big-Dipper-length) beyond the end-star Alkaid takes the eye to brilliant Arcturus. And several other clever lines and curves using Big Dipper stars have been noted.

Incidentally, we shouldn't mention the Big Dipper without saying something about the Little Dipper. The latter is the asterism formed by the main stars of Ursa Minor the Little Bear. The handle-end of the Little Dipper and tail-end of the Little Bear is Polaris. We will have much more to say about Ursa Minor, Polaris, pole stars of past and future, and positional arrangements of the two Dippers or Bears, in the June chapter—for June is the month when the Little Dipper is highest in the evening sky.

The Great Bear

Ursa Major has more stars with proper names than perhaps any other constellation. The Big Dipper is only the hindquarters and the tail of the Great Bear. And, by the way, bears do not have long tails—the one imagined for Ursa Major (and Ursa Minor) are unnatural in length. How could cultures who knew what bears looked like (the Native North Americans, even the ancient Greeks) have pictured these stars as a bear with such a ridiculously long tail? The answer in some cases is that they didn't. The early basic picturing by at least some Native Americans was that the bowl of the Big Dipper was the bear and the three stars of the tail were hunters following it. Only later did the concept of the long tail get picked up by almost every culture. For it is an odd fact that at least some of these stars of Ursa Major have been imagined to be a bear by a number of cultures widely separated in both miles and centuries. There has been serious speculation by respected authorities

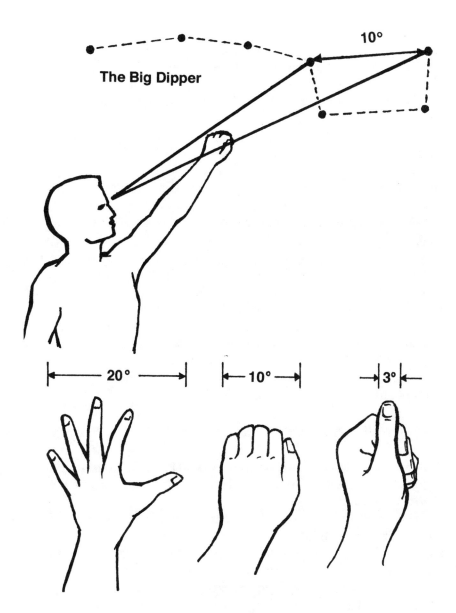

Figure 7.1. The Big Dipper as a check of angular measure—10° across the top of the bowl is about the same angular distance as the width of your fist held out at arm's length.

on astronomical history and lore that the idea of relating these stars to a bear goes back to a culture that lived late in the last Ice Age—ten or twelve thousand years ago! As tempting as this is to believe if you long for wonders and greater depths of meaning in life, I have always felt that the case for this Ice Age Ursa sounds pretty thin.

By the way, another reason why some cultures may have started calling these stars a bear is a confusion of language. Even in English there was a confusion between these stars being a great bear and a great bier. The Big Dipper, or its bowl, does looks like a wagon of some sort and indeed starting in the Middle Ages it was considered the wagon or chariot of the great ruler Charlemagne—in Britain "Charles' Wain."

We should note here that the sprawling patterns of stars which are joined with the Big Dipper to create the chest, head, and long legs of the Great Bear are interesting to trace. Many of these stars have proper names and some of them are, as we noted before, brighter than the dimmest star in the Big Dipper itself. Some of these stars have also been pictured in myths quite unconnected with bears, wagons, or the stars of the Big Dipper. For instance, the feet of Ursa Major today are imagined to be formed by three pairs of stars that appear as tight pairings to the naked eye. But in the Middle Ages the Arabs considered these pairs "the Three Leaps of the Gazelle"—the hoofprints left by a leaping antelope in a legend of theirs.

Deep-sky Objects in Ursa Major

A constellation as huge as Ursa Major would be likely to contain some impressive deep-sky objects for telescopic observers. And it does—especially galaxies.

Besides Mizar A and B (accompanied by Alcor), which we have already discussed, the Great Bear offers another of the sky's most outstanding double stars—Xi Ursae Majoris. It is also called Alula Australis, and is the southern member of the first and easternmost of the Three Leaps of the Gazelle. In the opening years of the twenty-first century, a 4-inch telescope on a good night is strong enough to show Xi as a 4.3 and 4.8 star only 1.8" apart.

But there are a few additonal facts which make Xi even more extraordinary. One is that the period of the secondary's orbit around the primary (or, rather, around their common center of gravity) is only sixty years. The position angle of the secondary changes noticeably every year. The separation varies from a maximum of 3″ (last seen in 1975) to a minimum of 0.8″ (last seen in 1992). It is about 1.8″ in the first few years of the twenty-first century and will only start opening really noticeably more widely around 2015. In 1780 the duplicity of Xi was discovered by Sir William Herschel, so around 2020 the secondary should complete its fourth fully observed orbit. In 1828, Xi was the very first double star to have its orbital period calculated. Another wonder is that this pair is only 25 light-years from Earth and that both stars are G0 stars. This means that the two are only a bit hotter than the Sun—and are probably both a little less luminous and large than the Sun (that's unusual in a naked-eye star let alone in the two components of a naked-eye star that is a fine double). The situation is complicated, however, by the fact that each of the two is actually itself a tight spectroscopic binary.

On the opposite side of Ursa Major is the finest galaxy pair in the heavens, a pair which is also a real odd couple. M81 and M82 are in a rather remote area of the Great Bear, but one way to get into their region is to draw a line diagonally through the Big Dipper's bowl from Phecda through Dubhe and extend it one more length of itself. M81 is a grand spiral which shines at about magnitude 7.0 and measures as much as 26′ by 14′. M82 lies just 38′ away and is a peculiar cigar-shaped galaxy. Though the total magnitude of M82 is about 8.4, considerably less than M81's, its dimensions are 11′ by 5′ and its surface brightness seems higher. I still recall my best view of M82, on a superb night with a 16-inch telescope. The mottling in it was amazing, the detail almost as rich as in long-exposure photographs from professional observatories.

It's likely that M82 was disrupted by M81 passing near it many millions of years ago. I've read different estimates of the distances to these galaxies but they are certainly close, forming the center of a group of galaxies not many millions of light-years beyond our own Local Group. No sight in the heavens is quite like this. You behold two of the sky's brightest galaxies as strikingly different-looking neighbors glowing ghostly and magnificent in the same field of view.

Another Ursa Major galaxy shines at magnitude 10.1 and can hardly be considered a showpiece in its itself. But M108 is located only 48′ from a different deep-sky object of greater interest and with which it forms another cosmic odd couple.

The neighbor of M108 (in the sky, not space) is M97—the Owl Nebula. This planetary nebula is located only 2.4° southeast of Merak, just outside the bottom of the Big Dipper's bowl. M108 lies almost right between M97 and Merak.

The Owl Nebula has often been listed as being magnitude 11.2— faint indeed for a nebula which measures almost 3' across. But I've never believed that the Owl was that faint visually and there are expert observers who agree. Steve O'Meara judges M97 to be magnitude 9.9. Of course, this still means the Owl is of low surface brightness. But I always remember my delight as a teenager in finding that on clear nights in my dark sky my small 4 1/4-inch reflector sufficed to show me M97 and M76 (another faint Messier object whose brightness is underrated).

With an 8-inch telescope, under very good conditions and with averted vision, one can certainly glimpse the two dark circular regions which are supposed to represent the eyes of the imagined owl. Considerably larger telescopes can show more detail.

Our final Ursa Major deep-sky object, M101, is a galaxy of great total brightness and very impressive size. This magnitude 7.7 object spans an area 27' by 26'. M101 is a face-on spiral but it has such large apparent size that there is a problem with surface brightness. Small telescopes show a smudge of light that is easily passed over. Larger amateur telescopes reveal a few brighter clumps of light here and there amid an amorphous glow. But under excellent conditions, large- or even medium-size amateur telescopes can begin to show hints of the spiral structure to a trained, patient observer. At least M101 is at a very convenient location. It forms an equilateral triangle with the two end stars of the Big Dipper's handle, in the direction away from the Big Dipper's bowl.

CANES VENATICI

A more intense face-on spiral galaxy is found on the other side of the Big Dipper's handle—in Canes Venatici.

Canes Venatici the Hunting Dogs was invented by Hevelius in 1687 from stars that formerly belonged to Ursa Major. Hevelius pictured the two dogs held on a leash by Bootes the Hersdsman but snapping at the heels of the dangerous Great Bear. The northern dog is named Asterion, the southern Chara. Chara is also the name of Beta Canum Venaticorum, the constellation's second-brightest star. The Alpha and brightest star of Canum Venaticorum is Cor Caroli— named by the physician of King Charles II of England in honor of King Charles I. The meaning of Cor Caroli is "Charles's heart." Cor Caroli is a striking and easy double star. The magnitude 2.9 primary has a 5.5 secondary a generous 20″ away from it. There is much controversy over the colors of these stars. See what you think! (Speaking of colors, one of the reddest 6th-magnitude stars in the sky is Y Canum Venaticorum, better known as "La Superba.")

Despite its several good double stars, the real deep-sky attractions of Canes Venatici are its galaxies and its great globular cluster. In the northern part of the constellation is a galaxy only about 3 1/2° southwest of the end star of the Big Dipper's handle. It is M51, but is also known as the Whirlpool Galaxy.

M51 is the most prominent of all the face-on spiral galaxies. M33 in Triangulum has a much greater total brightness but is spread out over a huge area and suffers from low surface brightness. Small or medium-size amateur telescopes cannot show the view of intense coiled spiral arms visible in the many stunning long-exposure images of M51. But in my experience even a telescope as small as a 6-inch can begin to provide hints of M51's spiral arms. This and larger telescopes also show an added attraction and peculiarity of M51: there is a sizable blob of glow outside the main spiral of M51. Indeed, large amateur telescopes (and, of course, photos) reveal that the blob seems to be attached to the end of one of the spiral arms of M51. This is the galaxy NGC 5195, which is actually a little farther than M51 but which astronomers believe has undergone interaction with the bigger galaxy.

M51 has a magnitude of 8.4 and covers an 11′ by 8′ area. NGC 5195 is 9.6 and 5′ by 4′. They are located only about 15 million light-years away from us. Together they offer what some observers consider the finest telescopic sight in the spring sky.

Yet Canes Venatici has other outstanding galaxies. M63 is the Sunflower Galaxy, a spiral which shows mottling and distinctive shape even in medium-size telescopes. It is slightly larger and almost as bright as M51. It lies 23.5 million light-years from Earth can be found about two-thirds of the way between the end star of the Big Dipper's handle and Cor Caroli. M94 and M106 are two other fairly open 8th-magnitude spirals in Canes Venatici which simply should not be missed. And if you like edge-on spirals then you should look up NGC 4631—the Humpback Whale Galaxy—and NGC 4244. These two shine at magnitude 9.3 and 10.2, respectively, and are about five and eight times longer in one dimension than the other.

Last but not least we come to the great globular cluster of Canes Venatici, the first really magnificent globular cluster of the year's progression. Few other globular clusters can compare with M3, rated at magnitude 5.9 by Steve O'Meara, and measuring about 10' across visually but much larger on photographs. A 6-inch telescope can begin to resolve many stars in this beautiful monster, but the total number has been estimated at about half a million. M3 is about 32,000 light-years away. Its only real competition among globular clusters in the north celestial hemisphere are M13 and M5 (autumn's M15 and, just south of the celestial equator, M2, are a little less bright and grand). M3 is probably observed less than the other great globulars because it lies in a rather star-poor area of eastern Canes Venatici, with no fairly bright stars nearby to act a as guide to it. You can locate it, however, a little more than halfway along the long sky journey from Cor Caroli to Arcturus.

CORVUS, CRATER, HYDRA, AND CRUX

From Canes Venatici we sail over the great galaxy wonderland of Coma Berenices and Virgo (we will explore them in the next chapter) and come to the southern regions of Corvus, Crater, and the middle of Hydra. In these regions, a few galaxies to rival even those of Canes Venatici can be found.

Before a telescopic tour, however, take a naked-eye look at the body of the Hydra, and at the Crow (Corvus) and Cup (Crater), which sit upon one of Hydra's last coils.

These stars of Hydra are mostly dim and Crater is a very faint constellation. But the modestly bright stars of Corvus are gathered into a fairly compact and quite geometrically interesting rhombus. Corvus also is located not far southwest of Spica and now, the time when it is approaching its peak height, it appears about due right of Spica for viewers at midnorthern latitudes. Even if a tree or building hides Spica, you may well notice this very discrete pattern of Corvus, standing out so well in such an otherwise star-poor region of heavens.

Corvus has been associated with several crows of mythology, including the one which was turned black as punishment by Apollo for the bird's disobedience and laziness. On the other hand, one of the most pleasant of Aesop's fables tells of a thirsty crow who couldn't reach down to the low water level in a cup to get a drink. The clever bird dropped stones into the cup until the level rose and it could quench its thirst. And indeed the fainter neighbor constellation of Corvus is none other than Crater the Cup.

The rear half of Hydra extends below Crater and Corvus and even onward. The end of this longest of all constellations has now finally risen up from the southeast horizon—by which time its head is past southwest.

By the way, if you set off straight south from Corvus and continue for about 40° of declination you come to an even more compact but much brighter and much more famous star pattern—the Southern Cross. The official name of the constellation is Crux, the Southern Cross. It contains two 1st-magnitude stars, one very bright 2nd-magnitude star, no even moderately bright star in its center (which disappoints some people), but also a large dark nebula called the Coalsack and the dazzling Kappa Crucis Cluster—an open cluster sometimes known as the Jewel Box. But of course the Southern Cross is forever below the horizon for most of this book's readers. If you live or travel to somewhere that is south of 30° N latitude, you can glimpse it. But you must be much farther south to see this famous pattern high enough to shine forth in all its glory.

More Corvus and Hydra

There is a fairly elusive but interesting planetary nebula (NGC 4361) near the middle of the Corvus pattern. There are also several fine double stars in the Crow. Delta Corvi—Algorab—is a wide double with a much fainter secondary that is nevertheless beautifully dissimilar in color from the primary. Struve 1669 is a 6.0 and 6.1 pair of yellow-white jewels that are 5″ apart. Double star expert James Mullaney notes that this pair is often encountered when one sweeps with the telescope from Delta Corvi to M104—M104 is the great Sombrero Galaxy, which lies just over the border in Virgo and so is discussed by us under the heading of Virgo, in next month's chapter.

Corvus does have an important galaxy of its own, even if it is much less bright than the Sombrero. It is NGC 4038-39, variously

Figure 7.2. The Southern Cross is in the center of this photograph by Akira Fujii, with the dark nebula called the Coalsack just to its lower left and the two brilliant stars Alpha Centauri and Beta Centauri at extreme left.

known as the Ring-Tail Galaxy or the Antennae. Though only magnitude 10.7 and 3' by 2' together, the Antennae turn out to be not one galaxy but two colliding and interacting galaxies. The appearance in amateur telescopes has been called "shrimp-shaped."

APRIL'S ANCIENT METEOR SHOWER

When you see some Lyrid meteors on an April night, you are seeing a meteor shower which has been observed at least as far back as 687 B.C.E. That is the earliest in history we have record of any of the major annual meteor showers. Yet the Lyrids have proven at least once in recent decades that they are still capable of remarkable surprises.

The Lyrids are believed to be derived from Comet Thatcher, which has an orbital period of 415 years and has been recorded only once so far as we know—at its most recent visit, back in 1861. These meteors can be seen from about April 16 to 25 but their peak comes on about April 22 each year. Their rates (numbers per hour) only stay above quarter maximum for about 2.3 days. In some of the final years of the twentieth century, ZHRs (Zenithal Hourly Rates—how many you'd see of the radiant was in your zenith and your limiting magnitude 5.5) of about 15–20 were observed for roughly eight to twelve hours, but in 2001 the Lyrids had ZHRs between about 20 and 35 for more than twenty-four hours. The most remarkable display of the Lyrids in recent decades, however, came in 1982 when ZHRs around 90 were observed for a short while. That outburst was in full view of North America and was a wonderful reward for faithful watchers of the old shower who were out that night.

The Lyrids are medium-swift and are rather bright on the average with some individual meteors that are spectacularly brilliant. Up to about a quarter of the Lyrids leave lingering glowing trains. The radiant of the Lyrids is located at about 18h4m and +34°, which is on the Lyra-Hercules border southwest of Vega. For viewers around 40° N, the radiant rises around 7:30 P.M., is respectably high by late evening, but doesn't reach its greatest height until around 4 A.M.

APRIL'S OBSERVING EXPERIENCE

For the past few decades, almost all the United States has used Daylight Saving Time (DT). As the twenty-first century gets underway, the only areas in the states which stay on Standard Time throughout the year are parts of Arizona and Indiana, and all of Hawaii. The rest of the United States remembers to "spring forward, fall back"—that is, to set the clocks forward one hour in spring and back one hour in fall. The date when the clock is set ahead in the United States is the first Sunday in April. To be precise, the change is to be made at 2 A.M. local time on that first Sunday—though of course most people simply change their clocks whenever they are ready to go to bed (if you turn in early for the night you might be setting your clocks on Saturday evening). In England, clocks are set one hour ahead (for "Summer Time") on the last Sunday in March.

What's important about the clock change for stargazers is that it makes nightfalls suddenly come even later by the clock than they were already getting to be. Instead of the Sun setting around 6:30 P.M. in early April, it is suddenly setting around 7:30 P.M. This gives the typical amateur astronomer fewer hours of darkness to use before bedtime. Of course, on the other end of the night, sunrise now comes an hour later. But this means about 6:40 A.M. instead of 5:40 A.M. and if you consider that morning twilight is getting too bright for most observations of stars by 6 A.M. Daylight Saving Time in early April, you can see that it is not much of a help to most stargazers (most of us are not accustomed to being wide awake and active outdoors between 5 and 6 A.M.—unless we are farmers, in which case we have morning chores!).

So in a sense, night is becoming a limited and therefore precious commodity in April—certainly one you have to wait longer to see. And when it comes you may find that for the first time in the year you have a new difficulty to contend with: insects. The good news is that many flying biters—various gnats—come out at dusk but disappear soon thereafter. Of course, it is also true that some April nights are still cold. Even in southern New Jersey I recall one year standing outside at 4 A.M. on April 21 with several inches of wet snow underfoot, a temperature of 31°F—and the sound of frogs in the distance! If you live farther north than I do, this may not be such a rare experience for you.

Make no mistake, however: April evenings are not all rain showers, bug bites, and snowfalls. The temperature is often just right, neither too hot nor too cold. Many pleasant nature smells abound—maybe the sweet scent of grass from your first mowing of the year if you observe from your yard. Mercury may be quavering in the dusk in a most unplanet-like fashion (because having a small disk and shining through low, unsteady air) and showing a bit of orange hue derived from a little haze and atmospheric absorption down low. Depending on where you live, leaves may be just returning to the trees—a welcome event by day but perhaps a greatly increased obstruction to your view of the night sky. (Even by night, however, masses of foliage can be a beautiful frame for a view of a group of constellations—a starscape is all the more beautiful in a treescape.) By day, swallows return if they haven't already and are twittering along with a rising-to-full-volume orchestra of other birdsong. But by night you may also hear some birds singing. In the pine-oak forests where I live, there always comes a nightfall in mid- to late April when I am watching the stars appear and hear breaking out, for the first time in the year, the hauntingly wild and beautiful call of newly returned whippoorwills.

Chapter 8

May

[Constellations covered: Coma Berenices, Boötes, Virgo,
Hydra (back section), Centaurus (briefly)]

I n May, the sun shines and zephyrs blow. Earth's north temperate
zone really is temperate this month. There are lots of deliciously
mild days, some with low humidity crowned with a deep blue vault of
heavens overhead. Such days are often followed by nights of unusually
dark skies in which stars therefore shine more brightly and more
numerously than the novice skywatcher could have imagined possible.

May is the month when even northern lands come into bloom.
(Memorial Day in the United States is held in late May because by
then there are, even in the northernmost states, flowers aplenty to put
on the graves of soldiers.) But May is also the month when the night
is abloom—for anyone with a fair-size telescope and a thirst for celestial wonders. For this is the month when the evening hours offer a
"meadow in the middle of the sky."

That phrase of the late, great astronomer and science popularizer
Carl Sagan was intended by him to refer to the sight of the Earth seen
hanging in the midst of space. A beautiful and effective use of the
expression! But what I mean by it here is something different. I mean
the region of the heavens which on May evenings looks relatively
devoid of naked-eye stars but which a telescope shows to be almost as
rich in fuzzy, subtly, and intricately patterned galaxies as any Earth
field is with flowers. Even if you don't have a telescope, there is the
lovely naked-eye sprinkling of a big, close star cluster in the northern

part of the meadow. And the meadow is bordered not by trees but by lines of both bright and not-so-bright naked-eye stars: the arching curves of the Big Dipper's handle, Boötes the Herdsman with brilliant Arcturus, Virgo the Virgin with bright Spica, and the trailing right triangle part of Leo the Lion.

Of course, we shouldn't neglect the rest of the sky on May evenings. Most of the May sky's very bright stars are quite low, either setting or rising. We are getting our last looks at Procyon, Capella, and at the two bright stars of Gemini, Castor and Pollux, which stand like twin sentinels over the final traces of evening twilight glow north of west. On another horizon, Deneb has come up low in the northeast to the lower left of Vega, which is now a third of the way up the sky. Another summer star, orange Antares, has just peeked above the southeast horizon. And yet another 1st-magnitude sun, Altair, will rise soon, a little north of east. Leo is still high in the west and the Big Dipper's handle hangs directly above Polaris in the north. The head and heart of Hydra the Water Serpent are still a third of the way up the southwest sky, but are snaking toward the horizon. Corvus the Crow sits on one of the middle wriggles of Hydra, and has just passed the meridian. So the sky as a whole is very interesting indeed on May evenings.

But in this chapter our featured constellations are the ones of the "meadow" plus others approaching or reaching the meridian: Coma Berenices, Boötes, Virgo, the tail of Hydra and, too low to be seen in its entirety by most of the world's population, Centaurus the Centaur. Our other topics include how to see pieces of Halley's Comet flaming in the May sky this and every year.

ARCTURUS AND SPICA

There are only three 1st-magnitude or brighter stars among the traditional spring constellations. One of them, the least bright of all 1st-magnitude suns, is Regulus, now halfway down the western sky as Leo declines. The other two stars are Spica, at almost precisely magnitude 1.0, and Arcturus at slightly brighter than magnitude 0.0 (magnitude –0.05).

Spica gets to the meridian before Arcturus, but if you look south in the middle of May evenings, you will see Spica partway up the south sky and Arcturus very high in the southeast. If you want to make sure that the bright stars you are seeing are really them, there is a fool-proof way. Take the Big Dipper's handle and extend its curve outward. Then all you have to do is "Make an arc to Arcturus, and drive a spike to Spica." Arcturus is about one Big-Dipper-length along the arc you extend from the handle. Driving a spike to Spica would mean taking a more straight line from Arcturus to Spica, but continuing the big graceful arc on past Arcturus will also bring you by a curve to Spica. There is even one more step you can take, and as far as I know I've invented it, or at least its saying: Continue the curve to Corvus.

Arcturus and Spica both have names relevant to their place in their constellation or their relation to other constellations. The name Arcturus means "bear guard" because the star and its constellation are supposed to be protecting all the other constellations from the fierce Ursa Major the Great Bear. The name Spica means "ear of wheat," because it is pictured as an ear of wheat being held by Virgo, who is usually identified with Ceres, the Roman goddess of growing things and the fruitful earth.

Arcturus and Spica may be partners of sorts, crossing the spring sky as the only very bright stars in their fairly dim constellations, the former star connectable by curve or spike to the latter. But they also offer many contrasts.

Spica is a spectroscopic binary. It is two hot B1 stars which are 260 light-years away and combine to burn about two thousand times brighter than our Sun (the primary is thought to emit about 80 percent of the total light). There are very slight (visually undetectable) variations in the brightness of Spica due both to actual pulsations of the primary star and to the two stars partially eclipsing each other as seen from our vantage point. The stars revolve around their common center of gravity in a period of just over four days, remarkably brief for giant stars. Spica A and B are calculated to be about eight and four times the diameter of our Sun, which would be about 7 and 3 1/2 million miles wide, and yet are separated by only 11 million miles. The masses of Spica A and B are perhaps eleven and seven times that of the Sun.

Arcturus is a single star (as far as we know), a K2 star only 36.5 light-years away which shines about 115 times brighter than the Sun. Arcturus is about four times more massive than the Sun but about twenty-five times bigger.

We will get to what sets Arcturus apart—well apart—from any of the other 1st-magnitude or brighter stars in a moment. But first let's compare the colors of Arcturus and Spica. Spica is said by many observers to be white. See, however, if you don't detect a hint of bluish in it with some optical aid. This blue might be exaggerated when Spica is near the yellow-white Moon, especially when it is about to pass behind the Moon's edge or has just emerged from it. (I observed one such occultation of Spica in which the star was easily visible in my 8 × 50 finderscope the last time I looked—about twenty minutes after sunrise.) Arcturus has a more unusual tint. Its B-V color index rating falls in a large gap between Pollux (lighter orange than Arcturus and also much fainter, so the color is less noticeable) and Aldebaran (deeper orange than Arcturus but considerably fainter, so its color is not necessarily more noticeable than that of Arcturus). Of course, the color a person's eye and brain combine to perceive and name varies from one indiviudal to another and is the result of a complex and somewhat mysterious process. A friend of mine once said that the color of Arcturus is "champagne shot with roses." See what you think. What is certain is that the star's hue, whatever it may be, is quite noticeable to the naked eye.

The Gypsy Star

Now let's get down to what really sets Arcturus apart from all of the other bright stars.

The space velocity of Arcturus—true speed through space in its orbit around the center of the galaxy—is twice as great as any other 1st-magnitude star and about three to ten times greater than most. Arcturus, it turns out, is not part of the main stream of traffic to which our Sun and almost all the naked-eye stars belong. It is not orbiting in the equatorial disk of the galaxy. It is going around the galactic center in an orbit considerably inclined to the plane of the galactic equator.

The amazing truth is that Arcturus just happens to be dropping through the equatorial disk at a point near our solar system. Most of our bright stars are in the flow of traffic with us and will remain fairly bright for at least a few million years in our sky. Arcturus is estimated to have first become visible to the naked eye only a few hundred thousand years ago. And a few hundred thousand years from now it will have dropped far enough away to once again be too faint for the naked eye. We just happen to be living in the time when Arcturus is within a light-year or so of its closest distance to us. Arcturus has a larger proper motion across the heavens than any of the very bright stars except for Alpha Centauri, which is about eight times closer to us. Arcturus is journeying at 2.28 arc-seconds per century toward the territory of the constellation Virgo.

Thus magnitude –0.05 Arcturus, the brightest star of the north celestial hemisphere, could be called a gypsy star, just passing through our neighborhood, destined for distant and exotic locales.

STARS OF BOÖTES AND VIRGO

Although Boötes and Virgo contain Arcturus and Spica, their other stars are only moderately bright. But that doesn't mean that the stars of Boötes and Virgo are uninteresting.

There is, for instance, one category in which Boötes may lead all other constellations: its number of bright, remarkable double stars that can be enjoyed by amateur astronomers.

Epsilon Boötis is the brightest Boötes double. It is called Izar but has also been called Pulcherrima—a name which means "the most beautiful." Although it is rather tight—the separation is 2.8"—the magnitudes are 2.5 and 4.9 and the color contrast is indeed one of the most beautiful in the heavens. Some say that the brighter K0 and dimmer A2 star look yellow and blue but the hues have also been called "pale orange, sea green."

Another gorgeously tinted pair in Boötes is Xi Boötis. This star system is only 21.9 light-years away and its two components, of magnitudes 4.7 and 7.0, revolve around each other in a period of only 150 years. During that time the separation varies from as much as 7" (as in 1977) to as little as 2" (as next happens in 2064). In the opening years of the twenty-first century the pair was about 6" apart and thus easy to split even in a 4-inch telescope. The colors have been described as orange and purple by some observers and yellow and reddish orange by others.

Zeta Boötis is a white and bright, equal pair (magnitudes 4.5, 4.6) but the pairing is close and getting closer throughout the first decades of the 21st century. It has a 125-year period and is currently closing from 1.2" in 1959 to 0.03" in 2021. Fortunately, stars of almost identical brightnesses are easier to split. At the start of the twenty-first century, the separation is about 0.8″—a real challenge for a 6-inch telescope. As the years pass, watch as it takes ever larger telescopes and more perfectly steady nights to split these two.

Other doubles in Boötes which are lovely and fairly easy include Mu Boötis (also called Alkalurops), Kappa Boötis, Pi Boötis, 39 Boötis, 44 Boötis, and Struve 1835. Double star authority James Mullaney also points out two examples of beautiful Boötes doubles which are sometimes not given in lists of best double stars because they are so wide. To see them appear fairly close together you do need low magnification. One of these systems is Delta Boötis, in which a 7.4 magnitude G yellow star is 105″ from a 3.5 magnitude G star. The other is Nu Boötis, in which the components are 15′ (1/4°) apart but are an orange and a blue-white star which both shine at magnitude 5.0.

What about the stars of Virgo? One of them is among the sky's most famous doubles. It is Gamma Virginis, also known as Porrima. (By the way, don't confuse Gamma Virginis with another famous double, Gamma Leonis.) Unfortunately for those of us living in the first decade of the twenty-first century, the Gamma Virginis pair is getting rapidly tighter and will soon be essentially too close together to split with any amateur telescope. The pair has a 171-year orbit and when most open (which last happened in 1919) are 6″ apart. But at the start of 2000 they were 1.5″ apart, start of 2001 1.3″ apart, and are now tightening even more swiftly. In 2008 the couple will reach a minimum separation of 0.4″ (theoretically, that could be split by a 12-inch or slightly smaller telescope even then but the "seeing" would have to be of a steadiness that is extremely rare in most climates). Thus, depending on when you are reading this book and how big your telescope is, you may or may not have any chance to split Porrima. Fortunately, the components separate very rapidly after 2008. They are back to 1″ apart by 2011 and 2″ in 2015. Gamma Virginis has a combined brightness of magnitude 2.2 with its components both 3.5, both F0 stars, and both white or slightly yellowish white. In fact, Robert Burnham Jr. wrote that this double looked like "twin headlamps of some celestial auto, approaching from deep space." The "deep space" here is not too deep: Gamma Virginis is only 39 light-years away. Incidentally, the duplicity of Porrima was discovered way back in 1718.

Gamma Virginis is one of several stars in Virgo which lie very close to either the celestial equator or the ecliptic—in the latter case, the stars have frequent encounters with planets. Beta Virginis is called Zavijava or Alaraph and is very near both the ecliptic and the celestial equator. Eta Virginis is called Zaniah and is closer to the ecliptic than Porrima. A star almost right on the celestial equator is Zeta Virginis, also known as Heze. Three other Virgo stars with proper names include Delta Virginis (Auva), Iota Virginis (Syrma), and Epsilon Virginis (Vindemiatrix).

Epsilon Virginis, which shines at magnitude 2.8, is a star of considerable importance to deep-sky observers. This star is the starting point for many forays into the heart of the richest of all bright galaxy clusters, the Virgo Cluster.

THE VIRGO GALAXY CLUSTER

The Virgo Cluster, or its heart, is sometimes known as "the Realm of the Galaxies." It was known as "the Realm of the Nebulae" before astronomers realized that all the fuzzy, patterned glows here were far beyond the bounds of our Milky Way Galaxy. Each of these objects is a galaxy in its own right. The Virgo Cluster is centered near the boundary of Virgo and Coma Berenices, spilling across part of both constellations, and so is sometimes called the Coma-Virgo Cluster. The Virgo Cluster is thought to be located between about 50 and 80 million light-years away and to contain about twenty-five hundred galaxies. There is a question, however, as to whether or not it is just the core of a larger assemblage of galaxy groups that together contains something like ten or twelve thousand galaxies. What's important to amateur astronomers is that roughly a hundred galaxies are visible with even medium-size telescopes within the approximately 12° by 10° region, and that in certain areas of the cluster the number of galaxies is especially amazing. Suppose you have an 8-inch telescope in rather good sky conditions or a smaller one in superb conditions. Then in a 1°-wide field centered on the galaxies M86 and M84 you may be able to glimspe about ten galaxies. With a slightly wider field of view and M86 and M84 at the southwest edge of it, you can behold about a dozen galaxies in a slightly zigzagging line or arc called Markarian's Chain.

Figure 8.1. The Virgo Galaxy Cluster contains the greatest gatherings of galaxies bright enough to see fairly well even in eight-inch telescopes. But this "Hubble Deep Field" image shows a plethora of incredibly dim galaxies all in just a tiny piece of sky near the Big Dipper. Objects as faint as 30th magnitude were recorded in this extremely long exposure with the Hubble Telescope.

The big problems are knowing where to start and knowing which galaxies you are seeing. Do understand that the Virgo Cluster is a real challenge to navigate for beginners or even for veterans if they are looking in fairly light-polluted skies. Of course, with an automated go-to telescope you can punch in the celestial coordinates of the galaxy you want and (if you've set up the telescope correctly) have the telescope place them right near the center of the field of view for you. But if you learn to "starhop" and navigate on your own through the Virgo Cluster, you will remember the major galaxies, find unexpected ones, see combinations of them which no book has identified, and get a rewarding direct grasp of the layout of the entire grand arrangement.

The Realm of the Galaxies

As we noted above, many observers begin their journey into the Virgo Cluster with Epsilon Virginis, the star which lies at the north end of the bent arm of Virgo formed by Gamma, Delta, and Epsilon. Once you've found 3rd-magnitude Epsilon, you must travel about 5° to its west to locate magnitude 4.9 Rho Virginis. There is no star near Rho that is comparable in brightness, but you must make certain that you have gone in the right direction from Epsilon. A check on whether you have Rho in the telescope is to note whether it is the bright star in the middle of a roughly north-south arc of stars (the star a small fraction of a degree to the north of Rho shines at 6.2, the one similarly near to the south of Rho is 6.9).

Now go 1.5° north from Rho and you will hopefully see at least two fuzzy patches of light. Most directly north of Rho Virginis is the magnitude 9.6 elliptical galaxy M59. Only 0.5° east of M59 is magnitude 8.8 elliptical galaxy M60. About 1° west and slightly north of M59 is the magnitude 9.6 spiral galaxy M58. Of these three galaxies, M60 is probably the most interesting. It is one of the most massive galaxies known, perhaps six times as big as our Milky Way. It has a faint companion galaxy you can glimpse with careful observation. Yet we can only touch upon the facts and appearances of such galaxies in the Virgo Cluster here, for there are just too many of them and too much to say!

What's most important is what kind of arrangement in the sky can be made of all the major Virgo Cluster galaxies, and how to navigate the lines of whatever patterns can be formed. I'm impressed with the fact that M60-M59-M58 form the southeast end of a 5° long line of seven Messier galaxies, or of a 9° long gentle arc (open to the northeast) of nine Messier galaxies. But ultimately I think the best way to organize the Virgo Cluster galaxies is by beginning with a coathanger pattern, which other deep-sky writers have noted. (By the way, be sure not to confuse this Virgo Cluster coathanger of galaxies with a small—but remarkable—asterism of the summer sky which is known as "the Coathanger"—see our September chapter for details.)

The accompanying map shows the coathanger arrangement of Virgo Cluster galaxies. Consult the map as you read the next few pages. Interestingly, the bottom line of the coathanger lies right on the much longer line we can draw from Epsilon Virginis to second-magnitude Beta Leonis (Denebola). As you can see on the map, the pair M60 and M59 (remember, only about 30′ apart) form one end of the pattern's

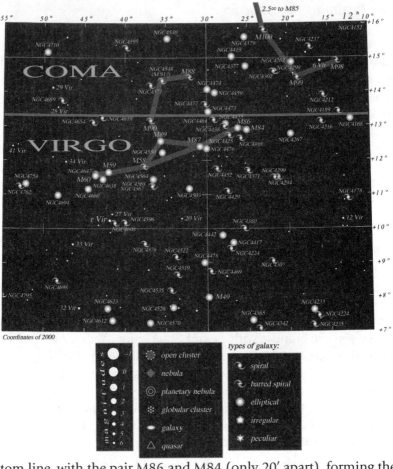

bottom line, with the pair M86 and M84 (only 20' apart), forming the other end of the bottom. In between these pairs, the galaxies M58 and M87 lie (well apart from each other) and form a wide base to a triangle whose top is M89. M89 begins the hook of the coathanger, which is continued with M90 (only 3/4° northeast of M89, so we could call them another pair). The hook reaches its top with M91 and curves back (southwest) with M88 forming the end of the hook.

The coathanger of galaxies includes ten of the Virgo Cluster galaxies that are Messier objects, all in a shape about 5° long and about 3° tall! Another pattern close to the northwest of this coathanger is formed by four more such galaxies: M99, M98, and M100 make a triangle, which hangs on a stem from the more northerly galaxy, M85. This entire triangle-on-a-stem lies in the constellation Coma Berenices, just north of its border with Virgo. The entire coathanger pattern of galaxies is in the constellation Virgo, just south of the border with Coma, except for the top of the coathanger, M91 and M88. (By the way, this is as good a time as any to mention that you may not find M91 on maps in older books—or you may find the desgination assigned to another galaxy. M91 is another of the "missing" Messier objects—no object exists in the location Messier recorded. But in recent years the prevailing opinion is that M91 is the galaxy otherwise known as NGC 4548, the magnitude 10.2 galaxy which is the top of our coathanger.)

Of the fourteen galaxies in these two patterns, the brightest of them in our sky and most massive in outer space is the nearly round elliptical M87. It shines at 8.6, and is 7' across. M87 is one of the most massive galaxies known in our universe. It contains about 800 billion times the Sun's mass and is believed to have a central black hole with a mass of about 3 billion Suns! M87 has long been known to have a strange luminous jet protruding from its center and is a mighty producer of radio and X-ray energy.

Other Galaxies and a Quasar

Do the coathanger and triangle-with-stem patterns locate for us all the Messier objects that are members of the Virgo Cluster? Not quite. Several degrees south of the coathanger is the second brightest galaxy in Virgo, the mighty 8.4 and 9' by 7' elliptical galaxy M49. It is 2 1/2° southwest of sixth-magnitude star 20 Virginis and has a 12.5 magnitude foreground star near its edge, which observers may wrongly think is a supernova in M49. Several degrees southwest of M49 is much dimmer but small and fairly prominent M61.

We should certainly note in passing that a few degrees southeast of M61 (and almost exactly 6° due south of M49) is 3C273—the brightest of all quasars (those mysterious superluminous objects thought to be the active cores of distant, early galaxies). You need to be a fairly experienced observer with a good finder chart to locate this 13th-magnitude (but variable in brightness) speck of light. Thirteenth magnitude seems quite faint to a beginner but it is reachable by a 6-inch telescope in good conditions and an even smaller telescope in superb conditions. And when you consider that 3C273 is 2 1/2 billion light-years away, you can begin to appreciate how prodigious quasar light must be (itself dwarfed by the radio output). Quasars are probably powered somehow by supermassive black holes. The light of the Virgo Cluster galaxies we see tonight has been traveling toward us since the last years of the dinosaurs. But 3C273—the closest quasar—is perhaps forty or fifty times more distant. The light from some quasars, visible in large amateur telescopes, has been traveling to us since before the Earth or even the Sun formed.

There is one last Messier galaxy far south of M49 and M61 and 3C273 which is believed to be an outlying member of the Virgo Cluster: M104. It is the brightest galaxy in Virgo. It is seen slightly tilted edgewise from our vantage point. It shows the most prominent true dust lane of any galaxy in the heavens. The lane can be readily glimpsed in 4- to 6-inch telescopes under good conditions. What is odd is M104's huge central bulge. This bulge, together with the dust lane bordering the bright equatorial disk and the galaxy's slight tilt, makes M104 look very much like a Mexican hat. That is why this spectacular object is widely known as the Sombrero Galaxy.

M104 is in Virgo but very close to the constellation's border with Corvus. The easiest (but not so easy) way to find it is to go 5 1/2° northeast of Eta Corvi, the star which, with Delta Corvi, marks the northeast corner of the Crow's pattern.

Before we leave the topic of the Virgo Cluster galaxies, I want to point out the fact that there are great numbers of the galaxy cluster's members that were missed by Messier. A map in Burnham's *Celestial Handbook* shows a roughly 9° by 5° section of sky in the Virgo Cluster. It shows galaxies down to 13th magnitude and there are about fifty of them. More impressively, while there are sixteen Virgo Cluster galaxies which are Messier objects and shine at 10.0 or brighter (two more—M91 and M98— are slightly fainter), there are nine galaxies at or brighter than the magnitude 1.0 limit in Virgo and Coma which are *not* Messier objects. They bear only NGC numbers. The brightest of them is NGC 4725 in Coma Berenices, which shines at 9.2 and measures 11' by 8'. The most striking of them is NGC 4565 in Coma Berenices. At magnitude 9.6 but measuring 16' by 3', it is the most dramatic of needle-thin edge-on galaxies (roughly an 8-inch telescope will begin to show its narrow dust lane).

I should add that some of the galaxies in the part of the sky occupied by the Virgo Cluster are not actually members of the cluster. NGC 4565, for instance, is much closer—"only" about 20 million light-years from us. There is even one Messier galaxy in Coma Berenices which is not part of the Virgo Cluster: M64, the Black-eye Galaxy. It lies well

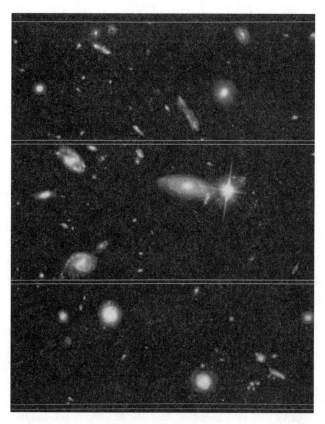

Figure 8.2. Although these selected areas of the "Hubble Deep Field" imaged by the Hubble Space Telescope are located in Ursa Major, they remind one of the rich concentrations of (much brighter) galaxies in the central part of the Virgo Cluster.

north of the galaxy cluster, less than a degree to the east-northeast of 35 Comae Berenices. Just 22 million light-years from Earth, M64 is a remarkable magnitude 8.5 spiral with a peculiar central region of dark dust cloud that is visible even in medium-size amateur telescopes.

COMA BERENICES AND THE COMA STAR CLUSTER

I started this chapter referring to what I called a "meadow in the middle of the sky." That, you remember, is the large area, mostly sparse of stars but rife with galaxies that is bounded by the Big Dipper, Boötes and Arcturus, Spica and Virgo, and the the right-triangle hindquarters of Leo the Lion. There is, however, already a giant asterism (unofficial pattern of stars) which encloses most of the "meadow." It is called the Diamond of Virgo, even though only one of its four stars is in Virgo. The Diamond is formed by: 0-magnitude Arcturus; 1st-magnitude Spica; 2nd-magnitude Denebola (Beta Leonis, tail end of Leo and eastern point of the Leo hindquarters tri-angle); and 3rd-magnitude Cor Caroli (the Alpha star of Canes Venatici, the fine double star we visited back in April).

Now, intertestingly, all the heart of the Virgo Cluster is wedged within the western angle of the Diamond, the one whose point is Denebola. Even if we don't count the very southerly M104 (which is not, for certain, a member of the Virgo Cluster), then the only one of the Virgo Cluster galaxies we have discussed which is to the south side of the line between Spica and Denebola is M61 (very slightly south of the line). It's also interesting that about two-thirds of the way along a line from Denebola to Arcturus (a line which bisects the Diamond) we can find Alpha Comae Berenices and, about 1° northeast of that magnitude 4.3 star, the 7.7-magnitude globular cluster M53.

The most fascinating object in Coma Berenices other than its galaxies, however, is a huge naked-eye star cluster which lies just about halfway along the line from Denebola to Cor Caroli. If your skies are quite dark you will not need to use those two bright stars to find it; you will simply notice it: a unique-looking collection of faint naked-eye stars, floating many degrees to the northeast of Denebola. Indeed, this star cluster was in ancient times considered the tuft on the end of Leo the Lion's long tail. In a moment, I'll relate the remarkable story of how the cluster went from being the tip of a lion's tail to the hair of a historical queen. But first let us consider this cluster's appearance and physical nature in more detail.

The cluster is technically named Melotte 111 but it is almost always called the Coma Cluster, short for the Coma Berenices Star Cluster. (There is also a distant Coma Galaxy Cluster, and the Virgo Galaxy Cluster is sometimes called the Virgo-Coma Galaxy Cluster, so one must be careful which names one uses.) The surrounding constellation named for the star cluster is called Coma Berenices, which means "Berenice's Hair." In a dark sky it is easy to see that the cluster itself looks like beau-tifully disheveled hair. For the Coma Cluster is an irregular scattering of a few dozen stars, at least five of them brighter than 5.5, and about a dozen or so brighter than 6.5, spread across an area about 5° in diameter. Magnitude 4.4 Gamma Comae Berenices is not a member of the cluster, though seemingly on its northern edge. Magnitude 4.8 star 12 Comae Berenices is the brightest true member. The distance to the Coma Cluster is only about 280 light-years, making it the third closest star cluster (only the Ursa Major Cluster and the Hyades are nearer).

How did this cluster go from being considered the Lion's tail tuft to being considered a lady's locks in ancient Hellenistic culture? The story is told that Berenice, the queen of Ptolemy III of Egypt (who

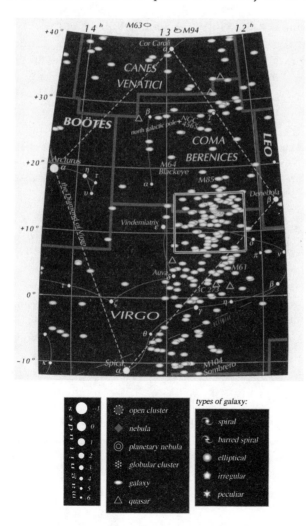

ruled from 246 to 221 B.C.E.), vowed that she would cut off her lovely "amber tresses" and offer them to the gods if the gods would permit her husband to return home from war safely. He did return safely and Berenice kept her promise, clipping her hair and having the locks placed in the temple. Soon, however, the hair mysteriously disappeared. What had happened? The court astronomer, Conon, explained that the locks of the queen had been accepted by the gods and been placed in the heavens to shine as a new constellation—the one which we today know as the Coma Star Cluster.

Inside the modern boundaries of the constellation Coma Berenices, within a degree of 30 Comae Berenices, is an important position in our sky: the North Galactic Pole. This point is as far as you can get from the galactic equator, which runs more or less down the center of the Milky Way band we observe in the sky. Evenings in April and May are when the Milky Way band is on the horizon, usually lost in the haze and horizon mist. When we look at this spot in Coma Berenices, we are looking perpendicular to the equatorial disk of the Milky Way in which we ourselves are located. This whole area of the heavens contains far less of the interstellar gas and dust which is thick in the equatorial plane. And that is why we have a "clear" view to see such distant assemblages of other galaxies as the Virgo Galaxy Cluster.

WONDERS OF HYDRA'S TAIL AND CENTAURUS

We complete our tour of May constellations by dropping far south from Coma Berenices to Hydra and Centaurus.

Hydra has deep-sky objects of considerable interest in its front two-thirds (see the chapters for March and April), but in its back third there is one quite lonely and quite southerly marvel which no one with a telescope should miss. M83 is a large, magnitude 7.6 face-on spiral galaxy. It is a magnificently detailed maelstrom of light as seen on photographs and some of that structure becomes visible in medium-size to large amateur telescopes. M83 is usually considered to be an Sc—loosely coiled normal spiral. It has, however, a strange element to its form which leads some astronomers to call the galaxy "three-branched." But other experts consider the feature to be some kind of distorted bar formation of stars, believing that M83 should be classified as a barred spiral. Perhaps this extra arm or branch was once a clearly recognizable bar that was disrupted by the passage of another galaxy (the best candidate is a galaxy that is now a few degrees from M83).

M83 suffers from getting only 20° high at best for observers at 40° N latitude. But much lower in the sky below M83—about 15° south of it—are two deep-sky objects which are perhaps even greater.

Both of these objects are in Centaurus the Centaur, most of which is too far south for a majority of this book's readership. One of the wondrous objects is a galaxy brighter and much bigger than M83 which seems to have the remnants of another galaxy smashing through it. It is NGC 5128, but is better known by the designation it is given as a powerful radio source—Centaurus A.

Just 4 1/2° south of Centaurus A is a globular cluster that is almost two magnitudes brighter than any of those which shine high in the skies of midnorthern latitudes. NGC 5139 appears as a bright fuzzy "star" to the naked eye and is best known by its original Greek-letter designation: Omega Centauri. It may be as many as a few million stars packed into an area of sky a little larger than the Moon. At magnitude 3 1/2, Omega Centauri is about a magnitude brighter than its closest competition among globular clusters, the even farther south 47 Tucanae. At the turn of the twenty-first century, however, astronomers raised the possibility that Omega Centauri is really not a globular cluster but a small, stripped-down galaxy captured by the Milky Way. Whatever it is, it is a magnificent sight if you can get it reasonably high above your horizon. I've seen it as a hazy, ill-defined but prominent ball in a 6-inch telescope even from about the latitude of New York City. At that latitude Omega Centauri appears just a few degrees above the horizon for a little while when at its peak in the due south.

There are two brilliant and important stars of Centaurus which are much farther south than Centaurus A and Omega Centauri. They are Alpha Centauri and Beta Centauri, respectively the third- and eleventh-brightest stars in the heavens. Alpha and Beta Centauri are 4 1/2° apart, the same separation as Castor and Pollux, but at magnitudes –0.3 and 0.6, these two put the far dimmer "twin stars" of Gemini to shame. Were it not for the fact that Alpha and Beta Centauri cannot be seen north of 30° N latitude, and are very low and dimmed even from places like southern Florida and southernmost Texas, their fame would be greater than almost any other stars. This is especially true because of two facts about Alpha Centauri: (1) it is perhaps the most spectacular double star system in the heavens; (2) it is the nearest of all star systems to our own.

Alpha Centauri lies 4.395 light-years from us. That's about 25 trillion (25,000,000,000,000) miles away, but almost twice as close as the closest other naked-eye star, Sirius. What's more, the point of light we see with the naked eye is actually a double, a 0.0 and 1.3 star taking eighty years to orbit each other. During that time, the separation changes dramatically, ranging from about 2″ to 22″. In the opening decades of the twenty-first century, the separation is closing rapidly, going from 12.0″ at the start of 2003 to 9.7″ in 2006, 7.4″ in 2009, 5.4″ in 2012, and 4.1″ in 2015. There is also a third star in the Alpha Centauri system, a red dwarf called Proxima Centauri. Proxima is an 11th-magnitude star almost 2° from the Alpha A and B pair. It is about one-sixth of a light-year from them, requiring something like a half million years to orbit them—and in our time it happens to be on the side of this orbit closer to us. The A and B star, when farthest apart, are separated by about the average distance between the Sun and Pluto, but when closest not much more than the distance between the Sun and Saturn. And it is the A star which piques our interest the most, for it is a G2 star like the Sun and is actually thought to be only a little larger, brighter, and more massive than the Sun. Alpha Centauri B is less massive and less luminous than Alpha Centauri A or the Sun, but it is a little bit bigger than both, for it is a cooler K-class star.

Alpha Centauri is sometimes called Toliman or Rigel Kentaurus (knee of the Centaur) and Beta Centauri is sometimes called Hadar or Agena. Beta may be close to Alpha in the sky but it is actually about 520 light-years from Earth and thus almost ten magnitudes or ten thousand times more luminous than Alpha.

MAY'S PIECES OF HALLEY'S COMET

As I write this book, more than fifteen years have passed since Halley's comet swept by the Earth and Sun in 1986. The next return is in 2061, almost sixty years away from the time of this writing. The nucleus of the comet will be a dark and inactive object beyond and below the orbits of Uranus and Neptune for decades to come. And yet twice in every year we have a chance to see mementos of the historic comet, actual bits and pieces of Halley dust streaking across our sky as meteors. The first chance each year comes with the flaming debris known as the Eta Aquarid meteor shower.

The Eta Aquarid meteors are caused by tiny bits of matter which were released in the dust tail of Halley's comet during previous returns. The debris is spread throroughly enough along the entire orbit of the comet for Earth to encounter some of it each year around the dates when Earth passes nearest the comet's orbit—even the years when the comet itself is 1 or 2 billion miles away in the outer solar system. The time we encounter the inward-passing part of Halley's orbit is for a few days around October 20; the time we encounter the outward-passing part of Halley's orbit is the first week of May—the time of the Eta Aquarids.

The radiant (source point) of the Eta Aquarids rises not long before dawn for observers at midnorthern latitudes, so for most of the world's population, this is the less plentiful of the two Halley meteor showers. Nevertheless, viewers around 40° N latitude and even considerably farther north have a chance of seeing at least a few of these very special meteors in the last hour of night on any date between about May 1 and May 10. Of course, as always, the chances of seeing meteors are greatly increased if there is no bright Moon in the sky and you are far away from city lights. Eta Aquarids shoot out of the southeast. They are very swift and about half of them leave persistent glowing trains.

MAY'S OBSERVING ENVIRONMENT

Even with Daylight Saving Time, morning twilights come quite early in May, even in the opening days of the month. Still, while you watch to catch sight of a Halley meteor or two, you may be hearing the choruses of different bird species beginning one by one. I remember a May dawn when I heard the clear song of a cardinal begin to flow in the dark woods while I watched Venus come so close to the crescent Moon that its rays seemed to touch the lunar edge. A study recently reported in *Science News* concluded that the amount of time a bird begins singing before sunrise depends upon the size of its eyes, with the largest-eyed birds starting earliest.

Of course, most of us are more likely to be out observing May's stars in the evening hours. But the fact remains that in May the music of birds, the blooming of flowers, the new foliage on trees are all at their peak, and that some of this vigor of song, scent, and sky-framing leafage works its way into our environment of the night. Mosquitoes and certain other biting insects can be a problem on May nights but only on the warmer, more humid ones which are precursors to summer. There are other May nights which can be among the clearest of the year, and with some of the most deliciously pleasant temperature to the air.

Not all but most of these comments about May are true in a majority of lands at midnorthern latitudes. A lot of the storminess of March and showers of April have given way to days of bright sunshine and cool but not freezing nights with which spring's plants can complete their growth. Of course, there are exceptions. There are a lot of lovely May nights in the central United States, but May is also the month when states like Oklahoma reach their yearly peaks of tornadic activity. On the other hand, the eastern United States benefits from the infamous Bermuda high of summer not yet getting established. There is also a weather pattern called an "omega block" which, when properly placed, often gives the East Coast spells of daytime humidities dropping even below 20 percent (rare for this part of the world) and deepest blue day sky followed by darkest star-filled night sky. I have seen the longest comet tails and many dimmest skyscapes on nights during these spells, which usually last several days and occur somewhere between late April and mid-May. In contrast, the omega block can bring many successive days of overcast to the East Coast in the second half of May.

On the shores of Delaware Bay near where I live, there is in May one of the world's greater spectacles of birds congregating—in timing with May's Full Moon. That Full Moon's high tide brings onto the beach for egg-laying the world's largest population of horseshoe crabs—not really crabs but strange living fossils with medically beneficial blue blood and a third, ultraviolet-sensing eye. Millions of birds arrive—in some cases from as far as the Southern Hemisphere—to feast on the eggs. Without this feast on their migrational route, a majority of the entire species of birds known as redknots might perish.

In short, May is a time for us to feel the vigor and vitality of all nature. In May, we can and should feel fully alive under the Sun by day and fully astir under stars at night.

Chapter 9

June

[Constellations covered: Corona Borealis; Hercules; Ophiuchus, Serpens Caput; Libra; Ursa Minor/Little Dipper, Draco]

J une is the month when summer weather hits the lands of the northern hemisphere. The summer solstice does not occur until three-quarters of the way through the month, but the summer weather patterns are usually well established weeks before then. What summer weather means for astronomers is often a major problem: hazy skies. Most of us associate haze with summer and many of us know that high humidity is a cause of haze. As we'll see, however, water vapor in the atmosphere is not the entire story and whether our day skies are milky and night skies washed-out with few stars depends also on another factor.

June brings the longest days of the year and also the longest twilights— in fact, if you observe as far north as 50° N latitude, there will be at least a few weeks when full night never comes. Twilight will reign from sunset to sunrise. But however dark your sky may or may not become, the arrangement of the starry heavens you'll be looking for is approximately the one shown on our map for June (in the back pocket of this book). Castor and Pollux are setting in the northwest. In the southeast at this hour, the teapot pattern of Sagittarius is sitting on the southeast horizon and is therefore dimmed by atmospheric absorption. But just west of rising Sagittarius and heaved up higher on the whole is striking Scorpius the Scorpion with its bright heart star Antares. About as high (still not very high) as Antares in

175

the east and northeast are other 1st-magnitude stars, Altair and Deneb. They form the base of a famous giant asterism; the Summer Triangle, whose brightest star, blue-white Vega, now sparkles well above the other two. The Big Dipper and Leo are still rather high in the northwest and west respectively, and Spica and Corvus are not too low in the south-southwest and southwest. Orange Arcturus is straight above (far above) Spica.

The prime, high constellations of June evenings are found in a strip of sky between brightest Arcturus and rival Vega, between Spica and Antares. This strip runs into the north sky, too. Our targets will include the northerly Little Dipper and Draco the Dragon; the big constellations Hercules and Ophiuchus, the former of which hosts the most famous of all globular clusters; Serpens Caput, the Serpent's Head, with its own superb globular cluster; Corona Borealis, the lovely crown of stellar gems which holds exciting variable stars; and Libra, the scales of Roman justice, once the claws of giant Scorpius. The subjects of this chapter will also include pole stars past, present, and future; highest Sun and lowest Full Moon; and much more. Let's begin our journey.

HERCULES AND THE NORTHERN CROWN

Strangely, none of our featured constellations of June are very bright. They are, however, rich with exciting deep-sky objects and especially rich in lore. Two of the best examples are the constellations of Hercules the Strongman and Corona Borealis the Northern Crown.

These constellations would not be easy for the beginner to find (particularly Hercules) were it not for a handy locational fact that is seldom mentioned: the key parts of both of these constellations lie directly on the line between the two brightest stars of June skies, Arcturus and Vega. The semicircle pattern of Corona Borealis is the first of the two that you encounter in scanning your eye from orange Arcturus to blue-white Vega. The second pattern is the important Keystone part of Hercules. The semicircle and Keystone are respectively about one-third and two-thirds of the way on the Arcturus-to-Vega line.

Corona Borealis

The semicircle is not hard to confirm once you are looking in approximately the right place, for it contains a magnitude 2.2 star. This is Alpha Coronae Borealis. It may be an outlying sun of the Ursa Major Cluster, whose core members are the middle five stars of the Big Dipper (see the chapter for April). This star is also known both as Alphecca and Gemma. The second name is appropriate because the star is indeed set like a gem in what is supposed to be the circlet of the Northern Crown—Gemma is from the Latin for "gem." Alphecca means "the broken," appropriate for the constellation because the imagined circlet of a crown is not complete.

The Northern Crown is that of the lady Ariadne. She is from Greek mythology, but is associated not with Hercules but rather with another hero, Theseus. She was the daughter of King Minos of Crete. Minos sent Athenian youths into the maze called the Labyrinth to meet their death at the hands of the dreaded Minotaur, who was half-man and half-bull. But Ariadne helped Theseus, giving him a spool of thread to let out behind him in the Labyrinth so that he could retrace his steps. Theseus was able to kill the literally bull-headed Minotaur and use the thread to find his way back out. Though Theseus had pledged his love to Ariadne, he treacherously left her stranded on an island on his way back to Athens. Fortunately the god Dionysius was moved by the sight of the sad woman and fell in love with her and made her his queen. When she died, Dionysius placed the crown of Ariadne in the heavens as the constellation we now see as Corona Borealis.

Corona Borealis is most famous for its amazing variable stars.

One is R Coronae Borealis, which variable star observers like to call R Cor Bor. Normally it appears as a 6th-magnitude star—the only star brighter than 6.5 within the half-circle of the Northern Crown pattern, or even within the complete circle or oval it would make if it were complete. R Cor Bor forms almost a perfect right triangle with Epsilon Coronae Borealis, the eastern end of the semicircle and Gamma Coronae Borealis, the star just east of Gemma in the semicircle. The amazing thing about R Cor Bor is that it suddenly and unexpectedly dims—sometimes to just seventh magnitude, other times to fifteenth, but usually to about twelfth! Then it begins to brighten, but slowly, the return usually taking a few months but sometimes even a year or two or longer. The dimming of the star requires only a few weeks, and in this day of Internet instant communication it is easy to get notified in time to watch the show. (See Sources of Information in the back of this book for how to receive AstroAlerts about important variable star happenings and other urgent celestial events.) Of course, it would be a special thrill to catch R Cor Bor's dimming by your own vigilance.

The leading theory to explain R Cor Bor's odd behavior is the idea that it may release carbon which condenses into a kind of sooty cloud that blocks most of the light from the star until the cloud disperses or falls back into the star. Whatever the cause of its behavior, R Cor Bor and the rare class of stars which resemble it have been called *reverse novas*, for they offer the opposite of a rapid brightening and slow dimming.

There is an even more spectacular variable star in Corona Borealis, though its shows are much more rare and harder to catch. This object is the brightest of a mysterious class of stars that are known as *recurrent novas*. The object is officially named T Coronae Borealis but is also known as "the Blaze Star."

T Cor Bor's normal state is a star of tenth magnitude. But on the night of May 12, 1866, this humble object was found to have become as bright as Gemma and then to reach magnitude 2.0, slightly outshining that star.

Was the Blaze Star a nova? Its fade was much faster than that of a nova. It faded about 1/2 magnitude a day after reaching maximum and dimmed below naked-eye visibility in just eight days. The star was back to its original 10th-magnitude state by June 7. Then, eighty years passed. On the night of February 9, 1946, T Cor Bor was observed to be at magnitude 3.2 but it was already fading and quite possibly its maximum had been as bright as in 1866, just missed by the world's observers. The Blaze Star faded almost exactly the same way it did before and shines even now at about tenth magnitude.

When might the Blaze Star flare up again? No one knows, but a few other recurrent novas have been discovered, much fainter than T Cor Bor but with more flareups observed (up to as many as five brightenings). Scientists think that perhaps all novas recur, it's just that these few recur in periods of less than a hundred years. There is evidence that T Cor Bor is an ultraclose double star in which matter from one sun is pulled onto another, periodically disrupting the latter. The Blaze Star increases its brightness by about 2,500 times during one of its outbursts. That is less than the 10,000 times or greater increases of traditional novas. But if T Cor Bor is as far away as believed then, its maximum luminosity is as great as a nova's—it simply doesn't fade to as dim a base state.

Receiving e-mail AstroAlerts about variable stars would seem to be almost crucial if you hope to catch the next performance of the Blaze Star while it is still near peak brightness. But of course you can always look for yourself if the night is clear. Is there a new naked-eye star perched only about a degree southeast of Epsilon, the eastern end of the Northern Crown's semicircle, and outshining Epsilon or perhaps even Gemma? If so, you have detected the latest phenomenal outburst of the Blaze Star!

Before you leave Corona Borealis, be sure to examine its easy, bright doubles, Zeta and Sigma. A very challenging duo is Eta Coronae Borealis. Located just 59 light-years from us, it is a pair of G2 stars which are both very similar to the Sun in size, mass, and luminosity. Their separation varies between 0.4″ and 1.1″ during a period of only forty-two years (the separation in space would vary from about that of the Sun and Saturn to about that of the Sun and Neptune). The similar brightness of the pair (magnitude 5.6 and 5.9) aids efforts at seeing both, but in 2000 the separation was only 0.7″ and will be 0.5″ in 2007 and 0.6″ in 2015. About as close during these years but even more difficult to see due to more unlike brightnesses (4.1 and 5.5) are the components of Gamma Coronae Borealis. Like Eta, this star should only be attempted on the steadiest of nights and only with a superb telescope with an 8- or 10-inch or larger aperature.

Hercules

Right up until the present, Hercules (Greek Heracles) has been the most famous hero of ancient Greek mythology. The legendary strongman's feats include those of his Twelve Labors, the first two of which involved slaying the Nemean Lion (sometimes identified with the constellation Leo) and the Lernean Hydra (often identified with the constellation Hydra). Yet the mythic hero himself became identified with a surprisingly dim, though large, pattern in the early summer sky. The pattern is that of a kneeling man and was known as the Kneeling One in pre-Greek cultures. It's even possible that this pattern was identified with the mythical Sumerian strongman, Gilgamesh, protagonist of the world's oldest piece of epic literature.

What Hercules lacks in naked-eye splendor it makes up for by its greatest deep-sky object: M13, the most observed and praised of all globular star clusters. To amateur astronomers with telescopes, summer is synonymous with M13.

To find M13 requires learning the Keyston asterism which forms the lower trunk of Hercules's upside-down body. As I noted a few pages back, the Keystone is located about two-thirds of the way from Arcturus to Vega. On a clear moonless country night, M13 appears to the unaided eye as a dim spot of hazy light on the west side of the Keystone. It is located about one-third of the way between Eta Herculis (northwest corner of the Keystone) and Zeta Herculis (the southwest corner). Binoculars or finderscope show the fuzzball bigger and two stars near to either side of it. A four-inch telescope swells the ball mightily and begins to show a few glittering pinpricks of individual stars. Larger telescopes reveal M13 as an ever more awesome object, with numerous individual stars and lines and curves of star concentrations sprawling from it like the arms or tentacles of a beneficent monster. At least hundreds of thousands of stars teem in this city of suns.

M13 is often listed as having a magnitude of 5.8 but it seems brighter, and Steve O'Meara estimates it as 5.3. It is certainly far smaller and dimmer than the great far-south clusters Omega Centauri and 47 Tucanae. But for people at midnorthern latitudes those objects are either not seen or not seen high enough for a proper view. M22 in Sagittarius may be greater than M13 in some ways, but it is fairly low in the skies for most North American and European observers and often spoiled by the thicker summer haze down low. By contrast, M13 passes almost right overhead for much of the world's population. Most observers think it is a bit more impressive than M3 and M5, its strongest rivals in the north celestial hemisphere.

While you are in Hercules, don't neglect trying to locate another globular cluster in the constellation—M92. Though smaller and less bright than M13, M92 is a fine object in its own right and looks considerably different than M13.

Hercules is also home to several striking and very interesting double stars. Among these are Kappa, Rho, and 95 Herculis, but two of the most interesting are Delta Herculis and Zeta Herculis. The latter

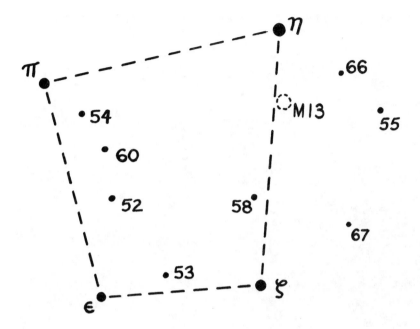

Figure 9.1. The Keystone asterism in Hercules, showing the position of M13 on the west side of the pattern. This is a good starfield for testing your limiting naked-eye magnitude at a site with a fairly dark or dark sky. The numbers with the stars are magnitudes with the decimal points removed (thus "54" means magnitude 5.4).

is the classic example of an "optical double"—two stars quite close in the sky (14″ apart) and yet at different distances, so actually many light-years apart. The components are magnitude 3.1 and 8.7. In contrast, Zeta Herculis is a true binary star and one of the finest of those whose components rapidly change positions. Just 35 light-years away, this "red and yellow" duo of magnitude 2.9 and 5.5 stars closes from 1.6″ (as in 1991) to 0.5″ (as in 2001)—so this is another pair that will be too close together for many amateur telescopes in the opening years of the twenty-first century.

Another double star in Hercules is Alpha Herculis, also known as Rasalgethi—Arabic for "head of the kneeling one." Rasalgethi does mark the head of Hercules. The brighter star in the Rasalgethi system is one of the hugest red giants known and varies semiregularly from about 3.1 to 3.9 in a period that merely averages ninety days. The secondary star shines at magnitude 5.4 only 5″ away. What is really spec-

Figure 9.2. M13 is generally regarded as the finest globular cluster that passes high in the sky at midnorthern latitudes. However, this Akira Fujii photograph shows the even more spectacular far-south object Omega Centauri, the richest and brightest of all globular clusters.

tacular here is the color contrast, for the brighter star is deep orange and the secondary has been described as vivid blue-green.

Rasalgethi, the head of the giant inverted Hercules, lies just a few degrees from the brighter head of a right-side-up giant constellation. The star is Rasalhague and the constellation is Ophiuchus.

OPHIUCHUS AND SERPENS CAPUT

Rasalhague is Arabic for "head of the snake-man" because Ophiuchus is the Serpent Bearer. This constellation separates the two halves of one official constellation, Serpens the Serpent. But the front (west part) of the snake is often called Serpens Caput ("the Serpent's Head" in Latin) and the back (east part) is often called Serpens Cauda ("the Serpent's

Tail"). In the picturing of Ophiuchus, the middle part of the serpent is being held by his waist-high hands, so that section of the serpent belongs officially to the constellation Ophiuchus the Serpent Bearer.

Who in mythology might this serpent bearer be? The common answer is Aesculapius, the god of medicine, whose snake-entwined staff is a symbol of the medical profession to this very day. Ophiuchus is a constellation centered on the celestial equator and lies halfway around the heavens from another such constellation which represents a giant with a name that begins with O: Orion. Interestingly, there is a story that Aesculapius (Ophiuchus) raised Orion from the dead after one of his two deaths—presumably the one caused by Scorpius the Scorpion. And Scorpius is the constellation below (south of) Ophiuchus, just possibly being tread on by the Serpent Bearer.

Ophiuchus has a huge pattern, a vaguely human outline of stars. Guy Ottewell says it looks like the Tin Man from *The Wizard of Oz*. But within that only moderately bright outline is a big area that lacks many stars. The area does include two magnitude 6 1/2 globular clusters only about 3° apart—M10 and M12. Other Messier globulars in Ophiuchus are less impressive, except for M62, in the southernmost parts of the constellation.

Ophiuchus offers a remarkable variety of deep-sky objects. Besides its globulars, it has a few big sprawling open star clusters. The best of these is surely IC 4665, located just a few degrees northeast of Beta Ophiuchi. Binoculars should be used to study this rather sparse but gigantic (almost degree-wide) cluster, whose total magnitude is listed as 4.2.

The area just east of Beta Ophiuchi holds several strange treasures. One of them is yet another fast-orbiting, rapidly changing, relatively near-to-Earth double star, 70 Ophiuchi. The 4.2 primary has been called yellow and the 6.0 secondary red. The pair revolve around each other in eighty-eight years but the minimum separation of 1.5" occurred in 1989, and the two were 3.7" apart by 2000 on their way to achieving a maximum separation of 6.8" in 2024. 70 Ophiuchi is the easternmost star in a tiny naked-eye triangle east of Beta and Gamma Ophiuchi. 67 Ophiuchi, the star just northwest of 70 Ophiuchi in the triangle, is a very luminous star about 2,000 light-years away, whereas the 70 Ophiuchi system is only 16.6 light-years from us (that is almost precisely the distance from us as brilliant Altair, by the way).

There are few naked-eye stars closer to us than 70 Ophiuchi, but only a few degrees northwest of it is a magnitude 9.5 star which is much closer. This is the famous Barnard's Star, the second closest of all known star systems after the Alpha Centauri system. Barnard's Star has a much greater velocity through space than Alpha Centauri, however, so it is the star of largest proper motion—that is, of greatest apparent movement across the celestial sphere. Barnard's Star moves more than 10″ a year. It is also getting closer to us. In the year 2015 its listed distance from us will have to be changed from 5.94 to 5.93 light-years! In the far future, Barnard's Star will come within 4 light-years of Earth and be about a magnitude brighter. This red dwarf star looks decidedly orange in a medium-size telescope, but to identify it you need a special finding chart.

Ophiuchus contains several other bright and beautiful doubles (39 Ophiuchi, 61 Ophiuchi . . .) but its two outstanding multiple stars—a triple star and quadruple star—are in the far south of the constellation, close to Scorpius. We will discuss them and the marvelous dark nebulas of southern Ophiuchus along with Scorpius in the next chapter.

Just west of Ophiuchus is Serpens Caput. Its brightest star is magnitude 2.6 Alpha Serpentis, also known as Unukalhai. Serpens Caput covers a much smaller area than the Serpent Bearer and has only a few notable deep-sky objects (Delta Serpentis is a 4.2 and 5.2 duo 4″ apart with subtle you-name-it hues). But one of Serpens Caput's deep-sky objects is the great globular cluster M5.

M5 shines at about magnitude 5.8. A 4-inch telescope can begin to show a first few individual stars glinting here and there in the overall mass of blaze. James Mullaney reports that in telescopes of about 10-inch aperture on superb nights some observers have reported M5's diameter as large as 27′. There are expert observers who even think M5 is grander than M13. One thing that M5 has in its favor is its proximity to the golden glow of magnitude 5.1 star 5 Serpentis (the star lies just 20′ from M5 and has a magnitude 10.1 companion 11″ from itself).

LIBRA

One of the dimmer constellations of the zodiac lies mostly southwest of Ophiuchus and Serpens Caput. It is Libra the Scales. Just as we are sometimes told to look between the bright stars of Gemini and Leo—

Pollux and Regulus—to locate Cancer, likewise we tend to search for Libra by looking between Spica in Virgo and Antares in Scorpius. The stars of Libra were in earlier times used to mark the big claws of Scorpius, but the Romans in the time of Julius Caesar changed them into Libra—the scales of Roman justice.

One surprise about dim Libra is that it has two stars which are actually moderately bright. Alpha and Beta Librae both shine at magnitude 2.6. The real reason for Libra's lack of prominence is the fact that its few modestly bright stars are scattered over a large area so that the best we can come up with for a main pattern is a too-big lopsided diamond. By the way, it must also be admitted that Libra is seriously lacking in exciting deep-sky objects. One of the few special sights is of a double star—Alpha Librae itself.

Alpha Librae consists of a magnitude 5.2 star that is 230" from a 2.8 star. That is enough separation to allow people with extremely sharp vision to see the fainter star with the naked eye under excellent conditions. On the other hand, any of us can detect Alpha^2 Librae (the dimmer component) with very slight optical aid (a weak finderscope or field glasses will do). These stars show common motion through space. They also lie very close to the ecliptic and thus are a prime target for conjunctions with the Moon and planets. In fact, back in 1972, observers in some parts of the world were able to witness the rare sight of Alpha^1 Librae being occulted (hidden) by Jupiter.

Beta Librae is another star of interest—due to its purported color. Some observers have seen a slight greenish hue to this star when viewed with optical aid. If so, it is the brightest single star to display this color. What do you think? Both the physiology and color vocabulary of different observers lead to different judgments about the subtle hues of stars.

Alpha and Beta Librae have arguably the most preposterous-sounding of stellar proper names, at least to English-speaking people. They are, respectively, Zubenelgenubi and Zubeneschamali—names derived from Arabic phrases meaning "the southern claw" and "the northern claw." This is a reference to the ancient placement of these stars in the claws of Scorpius.

By the way, south of Libra is the constellation Lupus the Wolf, which in star atlases is often shown being fought with a staff by its neighbor Centaurus the Centaur (half-man, half-horse). Lupus is mostly too far south to be of much interest for most of our readers.

But I can't resist mentioning that it was in this constellation that there occurred in 1006 a supernova almost as bright as a half Moon, the brightest we can confidently identify in the historical records.

URSA MINOR, THE LITTLE DIPPER, AND THE NORTH STAR

Now that we've toured the prime constellations of June evenings from Hercules and the Northern Crown near the crown (zenith) of the sky, down through Ophiuchus and Serpens Caput and Libra in the southern sky, it is time to turn our attention to what constellations are high in the north. One such constellation is Ursa Minor the Little Bear, whose main pattern is also known as the Little Dipper and contains one of the most famous of all stars: Polaris, the North Star.

We have already encountered the Little Dipper and Polaris in April, when we discussed the Big Dipper at its evening highest and how to use its Pointer stars to find Polaris. Now it is time to say more about the smaller asterism and constellation, and about the pole star.

We note that Ursa Minor is highest on June evenings. But admittedly, since the constellation is always visible as it makes its small circle around the north celestial pole, the time of year or time of night doesn't usually make a critical difference. Still, it is true that the two brightest stars of the Little Dipper's bowl are on June evenings about 30° to 35° higher than they are on December evenings. And for viewers at around 40° North latitude that means they are about 55° to 60° high as opposed to only 20° to 25° on December evenings (low enough to be considerably dimmed by atmosphere, haze, and light pollution—not to mention potentially hidden by trees or buildings). June evenings present us with the sight of the curved handle and bowl of the Little Dipper sticking straight up from Polaris, somewhat like a gymnast doing a handstand.

Polaris is at the end of the Little Dipper's handle and end of the Little Bear's unnaturally long tail. The rest of the handle and the stars of the bowl nearest to Polaris are quite dim, especially in comparison to the analogous stars in the Big Dipper. But the stars on the far side of the Little Dipper bowl from Polaris—Beta and Gamma Ursae Minoris—are of respectable brightness. Gamma, also known as Pherkad, shines at 3.0. Beta Ursae Minoris is also known as Kokab, and it glows at magnitude 2.1—only a tiny bit dimmer than Polaris itself. Kokab is a cool orange K star. Together, Kokab and Pherkad are called "the Guardians of the Pole" and, as we'll see in a moment, they have an interesting past.

First, however, let's consider Polaris. It has a magnitude 8.9 companion 18" from it which can be detected in rather small telescopes, but is easier in good medium-size telescopes and a joy to behold as an intense little spark near the main star's fire. Polaris is a distant and very luminous star and is actually a Cepheid variable—though one with so small a range that the variation is undetectable visually. Some scientists believed that Polaris's slight fluctuations in brightness (and pulsations) in size had finally ceased forever in 1994 but this proved not true.

Many beginners are shocked to learn that Polaris, the famous North Star or "Pole Star," shines no brighter than magnitude 2.0. That is similar in brightness to most of the stars of the Big Dipper but still only of second magnitude and much less bright than stars like Sirius and Vega (to mention a few of the several dozen which outshine Polaris). But of course it is Polaris's proximity to the north celestial pole that makes it really special. It is an excellent guide to true north—in our era of history, that is.

As explained in chapter 2, there is a slow wobble of Earth's axis called precession which changes the direction to the north celestial pole—and therefore which prominent star is closest to that pole. At the start of the twenty-first century, Polaris is within a degree of the north celestial pole. At the start of the twenty-second century—or, more precisely, 2102—Polaris will be at its nearest to the pole, just 27 1/2' from it. But for a number of millennia in the future, the Earth's axis will be pointing at various spots in the constellation Cepheus the King.

What's more interesting to most of us is where the pole has been in the past. About two thousand years ago, Kokab and Pherkad were much closer to the pole than Polaris was—though not nearly so close as Polaris now is. Thus in ancient Greek and Roman times it was these Guardians, or the entire Ursa Minor, that were watched by navigators. It was supposedly Thales of Miletus back around 600 B.C.E. who invented Ursa Minor—from some of the stars of a constellation we'll meet in a moment, Draco the Dragon. Our Little Bear and our Polaris were before then the wings of the Dragon!

The earliest pole star remembered in history is Draco's star

Thuban, which was close to the north celestial pole almost five thousand years ago and was followed by the ancient Egytians of that period. It takes close to twenty-six thousand years for the position of the pole to describe a large circle in the sky and return to its present location. On the opposite side of the circle from our own, the pole is moderately close to Vega. Thus Vega was a brilliant though not very precise North Star thirteen thousand years ago and will be again thirteen thousand years in the future.

DRACO

Draco the Dragon is an ancient and important constellation which seems to have become overlooked in recent decades. This is no doubt partly because Draco is only moderately bright as a whole and is a long constellation whose twists and turns require a bit of effort to follow. Add some light pollution to a typical observer's sky and most of Draco becomes less than prominent. There are also in Draco few galaxies, nebulas, and star clusters—the principal favorites of many of today's advanced amateur astronomers. Yet Draco has numerous fine double stars (rivaling even Boötes in this respect), one of the sky's few most dramatic planetary nebulas, and arguably the second best perfectly edge-on spiral galaxy in all the heavens. Draco also has one of the most important stars of our past and most exciting ones of our distant future.

Even before considering all these individual wonders, however, we should consider Draco as a whole and its position. The pattern of Draco is like a vast backward letter S or, on June evenings when it is high in the north, like a loopy, rounded capital letter N. The tail of the Dragon is the start of the N, which we can follow up right between the Big Dipper and Little Dipper. The first half of the N curves around to come down on the other side of the Little Dipper (thus the latter is half-encircled by the back half of Draco—don't forget that until about twenty-five hundred years ago the Little Dipper stars were the wings of the Dragon). The second part of the N—the front half of the Dragon—curls back on itself without enclosing any other constellation but ends with the most prominent part of the constellation: a little quadrangle of four stars. This compact and rather noticeable head of Draco is a magnitude scale in the sky, for it contains one star that is 2nd-magnitude, one that is 3rd-magnitude, one that is 4th-magnitude, and one that is 5th-magnitude. The head points toward Hercules but is also fairly near Vega, which can be taken as a 0 magnitude star in the scale with slightly farther-off Deneb as the 1st-magnitude example. Note that the head of Draco forms an equilateral triangle with Vega and the Keystone of Hercules (which includes M13) and that one of the wings of Cygnus the Swan points toward the Dragon's head.

The stars of Draco's head are, in order of brightness: Eltanin or Etamin (Gamma Draconis), Rastaban (Beta Draconis), Grumium (Zeta Draconis), and Kuma (Nu Draconis). Nu Draconis is a marvelous wide double—magnitudes 4.9 and 4.9 and both white A5 stars, 62″ apart. They can be split with even the slightest magnification (there is even the chance that this is the ultimate star test of very sharp naked eyes, though I'm not aware that anyone has ever split Nu Draconis without optical aid). By the way, a little star which extends the pattern of Draco's head symmetrically to the west is also a pair of same-brightness, same-spectral-class suns: this is Mu Draconis, also known as Arrakis and in this case magnitude 5.7 F7 stars—but only 2″ apart.

Of course, far more prominent than Nu and Mu Draconis is Eltanin. It shines at magnitude 2.2 with a rich deep orange hue comparable in binoculars to that of Betelgeuse and Antares. Consider it the eye of the Dragon. If you do, then I can tell you that this eye is brightening as the star nears us in the next few million years. At a far future date Eltanin should be the brightest star visible from Earth, a kindled "ember" brighter than Betelgeuse now appears.

Now you may be wondering where the Alpha star of Draco is, since it is not 2nd-magnitude Eltanin or any of the other stars of the head. The answer is that Alpha Draconis is a star near the tail end of the Dragon which shines at a modest 3.7. The reason it was accorded the alpha designation is no doubt because of its tremendously important past. For Alpha Draconis is indeed Thuban, the star that was very near the north celestial pole almost five thousand years ago. Thuban is actually still easy to find, for it is the star which lies about halfway between the bend in the Big Dipper's handle (marked by Mizar) and the bowl of the Little Dipper (marked in part by the Guardians of the Pole). By the way, the next star east from Thuban in the body of Draco is Iota Draconis (Edasich—Draco certainly does have a lot of named stars!) and several degrees from Iota are two 10th-magnitude galaxies. The more interesting is NGC 5907, a 12' by 2' edge-on spiral which many observers feel is surpassed in its class only by NGC 4565 in Coma Berenices.

Among the beautiful double and multiple stars of Draco other than Kuma and Arrakis are Phi Draconis (4.9 and 6.1, 30" apart), 17–16 Draconis (5.4, 6.4, 5.5 with separation of 3" between first two and 90" between first and third), and 41–40 Draconis (5.7, 6.1, 7.5 with separation of 19" between the first two and 222" between the first and third).

But there is one more deep-sky object in Draco to mention and one more positional attribute of the constellation to describe. Within the coil of the front half of Draco lies a pole of more lasting importance than the particular north celestial pole of any given era of history: the north ecliptic pole. This is the north pole of Earth's orbit (and nearly that of the whole solar system). It is the center of the giant circle which the north end of Earth's axis delineates in the heavens due to precession over the course of 25,800 years. Thus, like the dragon wrapped around the tree holding the Golden Fleece or the dragon wrapped around the tree bearing the golden apples of the Hesperides, Draco is curled around an axis of great importance. And what deep-sky object is extremely close to this north ecliptic pole? NGC 6543, the 9th-magnitude planetary nebula called "the Cat's Eye Nebula," prominent in small telescopes, and one of the most spectacular-looking objects seen on photographs by observatories and images by the Hubble Space Telescope.

JUNE'S HIGH SUN AND LOW FULL MOON

The June solstice occurs around June 21. At that time, the Sun reaches its most northerly point in the heavens, marking the start of summer in Earth's northern hemisphere and of winter in its southern hemisphere. North of the equator, this is the longest day of the year, the Sun staying above the horizon just over fifteen hours at 40° N latitude. Farther north, the day lasts progressively longer, finally achieving twenty-four hours at the Arctic Circle. This is also the time of year that twilights are longest, about two hours of dusk and two hours of dawn occurring around 40° N compared to only 1 1/2 hours

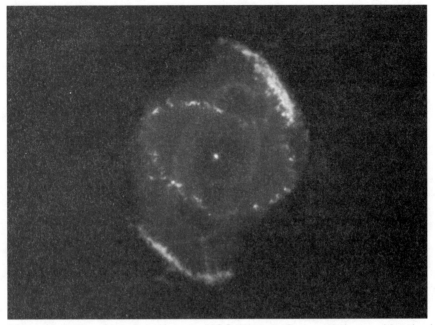

Figure 9.3. The Cat's-Eye Nebula, NGC 6543, in Draco, as imaged by the Hubble Space Telescope.

of each around the equinoxes. And around 50° N—for instance, in England—there is for several weeks no true night, the latest dusk flowing uninterruptedly into the earliest dawn.

At the summer solstice, the northern hemisphere of Earth is tilted its most toward the Sun—though we should remember that this orientation occurs because the Earth now reaches a certain place in its yearly orbit, not from any change in the direction of Earth's axis relative to the stars. (The direction of Earth's axis remains constant enough during a year, requiring, as we just discussed, thousands of years to change substantially.)

The summer solstice is when the Sun is highest in the sky for viewers at mid-northern latitudes. But how high? Since Earth's axis is tilted 23 1/2° to the plane of its orbit, it is at 23 1/2° latitude that the Sun appears exactly overhead on the summer solstice. This latitude line on Earth is called the Tropic of Cancer (its counterpart in the southern hemisphere is the Tropic of Capricorn) because when it was named a few thousand years ago the Sun made its trope—its "turning point" back south—in the constellation Cancer. Precession of Earth's axis—and therefore "precession of the equinoxes" and with them the solstices—has brought the Sun's far north point to Gemini. The Tropic of Cancer passes south of southernmost Florida and Texas, so no spot in the continental United States ever quite sees the Sun at the zenith. The tropic line passes right over the southern tip of the Baja peninsula (shadows should shorten and disappear at true noon there this day) and slightly north of Hawaii—so at summer solstice the Sun passes slightly north of the zenith as seen from Hawaii.

The Full Moon is in the opposite direction in the sky from the Sun. This means that the Full Moon of June—or, more precisely, the one occurring nearest the summer solstice—is the lowest of the year. June is famous for weddings, and Guy Ottewell has suggested that the word "honeymoon" may derive from the color of the June Full Moon—which is often a rich deep gold due to shining through summer haze low in the sky.

JUNE'S OBSERVING EXPERIENCE

Haze is a fact of life for many of us on many summer nights, especially in June. Instead of seeing a deep blue sky in the afternoon, we often see a milky, brightly glaring one. Instead of seeing a dark sky with lots of stars at night, we often see a brightish, washed-out sky with few stars prominent to the naked eye.

Hazy summer nights are a part of most observers' experience and some would say that all experiences are valuable and should be appreciated. Realistically, however, these nights are not good for many types of stellar and deep-sky observation. One exception, though, is double-star viewing. Haze gets established in air masses which are stagnant from lack of movement. Such a still atmosphere, leading to steady images, is just what an observer needs to split close double stars—and also to see fine details on the Moon and planets. Of course, too much haze will make it difficult to see all but the brighter double stars and will obscure lunar and planetary detail. But nights of a little haze can be excellent ones for viewing the many double stars of Boötes, Draco, Hercules, Ophiuchus, and other constellations which just happen to be at or near their highest on June evenings.

It is well known among weather buffs that warm air can hold more moisture than cold air can. But this is not the entire reason that many of us have hazy skies in summer. In reality, another key factor is manmade—not only the production of smog, mostly from car fumes around big cities, but more pervasively, the vast quantities of sulfur dioxide gas being released by coal-fired power plants. The sulfuric acid haze derived from *combining* this gas with large amounts of water vapor (summer humidity) is a major culprit in making skies hazy in large parts of the industrial countries of the world that are near the plants or downwind from them (for instance, much of the eastern United States). Thus, sometimes you will notice that the humidity is high on a summer day, but the prevailing wind is not from the direction of the coal plants so that the sky looks a deep "tropical" blue—and the night can be rich with stars.

Incidentally, in the United States the haziest season of the year up until the middle of the twentieth century was often winter—the time when wood was burned to heat many homes. The increased use of oil and gas for heating, combined with the increase of coal-fired energy plants, changed the season of most haziness.

Most of us long for the rare passages of cold fronts in June to clear out the air and give us richly starred skies. But maybe these hot and hazy nights can be ones to treasure in their own way. While you are looking at double stars and any available Moon or planets, you may find yourself surrounded by the year's first and maybe greatest "galaxies" of fireflies, briefly shining almost exactly like stars when seen high in the distance. On a first humid night, the sound of cicadas freshly erupted from the earth may join the choruses of the night loudly. And you may see quite a few stars overhead while lazy flashes of heat lightning play idly on one or more of your horizons.

Chapter 10

July

[Constellations covered: Lyra, Aquila, Scorpius,
southern Ophiuchus]

There are, in my opinion, two quintessential stars of summer. One glitters blue-white, the other golden orange. One shines high in a tiny constellation, a constellation which passes virtually overhead for much of the world's population. The other glows low in a huge constellation, a constellation which creeps down near the south horizon for much of the world's population. The high, bluish star is Vega and it is often associated with the two other 1st-magnitude stars of the "Summer Triangle" asterism—Altair and Deneb. The low, almost ruddy star is Antares and it is usually associated only with its brilliant and giant constellation, Scorpius the Scorpion.

When at nightfall Vega is very high and Antares is as high as it gets—though still low, especially as seen from places as far north as southern Canada and England—summer is at its peak. Where I live, wild blueberries and blackberries are then ripe and ready for the picking, the nights begin to fill with the year's largest total number of meteors, and other signs of plenty and fulfillment abound.

Among other culminations, July usually brings the culmination of the year's trend of rising temperatures in Earth's northern hemisphere. The highest Sun and longest days are around the summer solstice, which occurs in June. But Earth's vast weather systems lag behind by several weeks. Just as the coldest weather is on the average in January about a month after the winter solstice, the hottest weather is on the average in July about a month after the summer solstice. And just as a person may be surprised that the Earth is nearest to the Sun (perihelion) in early January, so, too, may a person be surprised that the Earth is farthest from the Sun in early July (aphelion). The variations in Earth's distance from the Sun during the year are far less important to climate than how much a given hemisphere of Earth is tilted toward (summer) or away (winter) from the Sun.

Last month, we examined many constellations, none of which was very bright, but most of which contained quite a few fascinating deep-sky objects (mostly double stars, but also several of the sky's premier globular clusters, and some other interesting entities). In this month of July, however, we will focus on mostly just three constellations (plus a rich part of Ophiuchus we didn't cover last month). One of these three, Aquila the Eagle, has few deep-sky objects but features the bright and marvelous Altair. The other two constellations are those whose chief lights are Vega and Antares: Lyra the Lyre and Scorpius the Scorpion. And it is to them that we are devoting most of this chapter. They are simply brimming with wonderful sights. Scorpius contains more 2nd- and 3rd-magnitude stars than any other constellation. It also offers several of the sky's best globular *and* open clusters, amazing double stars, a recently dramatic variable star, and much more. Lyra is tiny and yet offers not just Vega but one of the sky's most prominently performing variable stars and both the most famous planetary nebula and the most famous (and spectacular) double-double star in all the heavens.

Add to these stellar marvels a few more special July sights like the plentiful Delta Aquarid meteor shower and you begin to appreciate why July is as much of a peak month in the heavens as it is on earth.

VEGA AND LYRA

I like to call Vega the queen star of summer and also "the sapphire of summer." At magnitude 0.03 it is the brightest star of the summer constellations. It is the brilliant star which passes near the sky's zenith for far more of the world's population than any other. (Vega's declination is +38°47', so it goes within a few degrees of overhead for anyone who lives near 40° North latitude.) The blue tint in its light is pretty easy to notice on most nights. A friend of mine has referred to Vega as "the Sirius of Summer."

This comparison of Vega to Sirius is interesting, for the two are similar in ways that go beyond both being blue-white and the brightest star of their respective seasons. Both Vega and Sirius belong to spectral type A, often known as "Sirian stars." According to some authorities, they even belong to the same precise spectral class, A0. Sirius is the star of greatest apparent brightness that is visible in the night skies of Earth. But Vega is about three times farther away, so it is almost a magnitude brighter in true brightness. Still, Vega is one of our closest bright stellar neighbors, at a distance of only 25.3 light-years. It is a star about two and one half times wider than our Sun with about fifty times the Sun's luminosity (Sirius is more than one and one half times wider than our Sun with more than twenty times the Sun's luminosity).

What is an interesting coincidence is that these two most luminous stars within about 25 light-years of Earth—Vega and Sirius—are, respectively, close to the point in space that our solar system is heading toward and close to the point in space that it is heading away from. To be more precise, these two points are the Apex of the Sun's Way (ahead of us) and the Antapex of the Sun's Way (behind us) and they are determined in relation to the motions of the stars in the Sun's "local" neighborhood. The Antapex—also known as "the Sun's Quit"—was discussed in our January chapter. The Apex is located at roughly 18h4m and +30°, which is actually just over the border from Vega's constellation Lyra, in the constellation Hercules near the stars Nu and Xi Herculis. But this determination is approximate and it is perhaps better to say that our direction is roughly toward Vega: Vega is the close, bright, blue-white star out our front window and Sirius the close, bright, blue-white star out the back window of our solar system. Of course, this is not to say that we were once extremely near Sirius or that we will ever, in the future, get extremely near Vega. For Vega and Sirius are moving, too. They merely mark, for now, the directions our Sun respectively came from and is going to.

Does Vega have a white dwarf companion like Sirius (and Procyon)? No, but there are two magnitude 9.5 "optical" companions— stars merely on about the same line of sight as Vega, not physically related. The companion much closer to Vega is 63" away, and in a well-collimated medium-size telescope stands out like a tiny piece of intense glitter in the almost ten-thousand-times-greater wash of trembling blue-white radiance of Vega itself.

Vega has been called the "Arc-light of the Sky" and has played a role in a number of science-fiction and fantasy tales, including H. P. Lovecraft's macabre work *The Dream-Quest of Unknown Kadath* and, more recently, Carl Sagan's novel (and later movie) *Contact*. (In *Contact*, Sagan imagined that the first messages from extraterrestrial intelligence came from the vicinity of Vega.) Vega and another bright star at almost the same distance from Earth (Fomalhaut) were shown by the IRAS satellite in the early 1980s to possess clouds of particles—likely belts of comets or other such material—surrounding them at distances greater than that of Pluto from the Sun.

Vega has also been called "the Harp Star," for it shines in the constellation of Lyra the Lyre, and the lyre is an ancient musical instrument which resembles the harp. Lyra represents the seven-stringed lyre of Hermes, later made famous by the greatest of all musicians of legend, Orpheus. The first lyre or harp was supposed to have been created by Hermes from a tortoise shell. That is not, however, the origin of the proper name of Beta Lyrae—"Sheliak," which comes from an Arabic name for the whole constellation that means "the harp." But Gamma Lyrae is called Sulafat, a name which is derived from the Arabic *al-sulahfat*—"the tortoise."

What about the name Vega? This has been used for about a thousand years and is from an abbreviation of the early Arabic name, which meant "the Swooping Eagle (or Vulture)." The name was also used to apply to the entire constellation Lyra. Thus all three of the constellations which bear the three bright stars of the Summer Triangle have been regarded as birds (the two constellations other than Lyra—Aquila the Eagle and Cygnus the Swan—are still known as birds today).

The Double Double

As Guy Ottewell has noted, the compact constellation Lyra looks more like a little rattle than a lyre, a rattle with Vega as the handle. But there is a star of modest naked-eye brightness only about 1 1/2° from Vega which would then also be part of the handle (perhaps a little projection on the handle). This is the star Epsilon Lyrae, but it is better known to amateur astronomers everywhere as "the Double Double."

A look at Epsilon with a keen eye on a steady night shows it to appear somewhat elongated or, if your eyes are extremely sharp, you

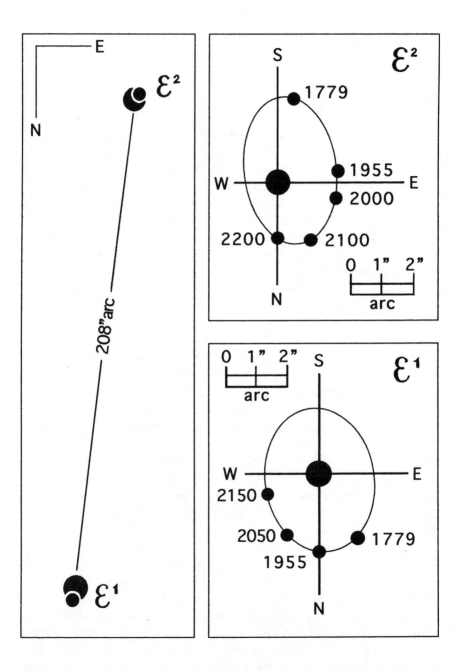

Figure 10.1. Diagrams of the Double Double, the Epsilon Lyrae system.

may be able to tell that it is really two stars. Most of us mortals will need to use a little magnification—though very little indeed—to plainly separate Epsilon into its two white component stars: a 4.6 and a 4.7 star 3 1/2′ apart. But there is more to be seen. You will usually need a good 4-inch or 6-inch telescope at 100× to 125× (though 2.4-inch and 70× will suffice under the best conditions) to have a revelation: each of these two stars itself splits into two similarly bright stars. The slightly wider pair (2.6″ apart) are magnitude 5.0 and 6.1, with spectral classes A2 and A4, respectively. The slightly tighter pair (2.3″ apart) are 5.2 and 5.5, and are A3 and A5 stars. The angular distances between the two stars in each pair are changing only very slowly (over the course of hundreds of years). The two pairs are traveling through space together. By the way, note that the axis of one pair (that is, the straight line drawn from one component through the other component) is at almost a right angle to the axis of the other pair.

There are other fine double stars in Lyra. There is the extremely wide (10 1/2′—a separation similar to Mizar-Alcor) orange 4.6 and blue-white 5.6 pair Delta^2 and Delta^1 Lyrae, a pair surrounded by a sparse but charming star cluster. There is Zeta Lyrae—a 4.3 and 5.9 star of different hues an easy 44″ apart. But it's a fainter double I'd like to mention, partly because it is still not well known and partly because of its relationship to Epsilon Lyrae. I'm referring to Struve 2470/2474— the star system nicknamed "the Double Double's Double"!

Struve 2470 and 2474 are about 10′ apart (three times the separation of Epsilon-1 from Epsilon-2 Lyrae). Struve 2470 breaks into a pair of 6.6 and 8.6 stars that are 13.4″ apart; Struve 2474 breaks into a pair of 6.7 and 8.8 stars that are 18.5″ apart. In Struve 2470/2474 the axes of the two pairs are almost exactly aligned (point in the same direction, have the same position angle)—in wonderful contrast to Epsilon Lyrae in which, as we noted above, the two pairs are at right angles to each other.

The components of Struve 2470/2474 are as much as two to three magnitudes dimmer than those of Epsilon Lyrae, and considerably farther apart. But the two systems lie within the same small constellation, in fact only about 8° apart (I can observe them both as single pairs in one field-of-view with my ultrawide-angle binoculars!). And the relative rarity of such systems anywhere in the heavens does make Struve 2470/2474 seem reminiscent of Epsilon to the observer. The former does deserve to be nicknamed "the Double Double's Double." Now if we could just find a third system in or near Lyra to brand with an even more outrageous appellation: "the Double Double's Double's Double"!

A final set of surprises arises when we consider the distances and luminosities of the Double Double and its double. The Epsilon Lyrae system is located about 161 light-years away from us. Struve 2474 is almost exactly as far! What's more, its G spectral type pair should be only a little brighter in true brightness than our G spectral type Sun. But what about Struve 2470? It turns out that Struve 2470 and 2474 are an optical double—the former lies roughly 1,900 light-years away from Earth.

The Ring Nebula and Beta Lyrae

The Double Double is one of the most familiar and favorite sights of summer for amateur astronomers. But even more famous and popular—indeed maybe the most popular deep-sky object of summer along with the great M13 globular cluster of Hercules—is the planetary nebula which sits conveniently a little more than half of the way along the 2°-long line between Gamma and Beta Lyrae (the southern stars of the Lyra pattern). This object is M57, the Ring Nebula.

M57 doesn't have the greatest total brightness of any planetary nebula (the huge and therefore low-surface-brightness Helix Nebula does). It doesn't even have the best combination of overall brightness and size; the Dumbbell Nebula does. But the Ring Nebula is a fairly close runner-up in the second of these catgeories, has its incredibly easy-to-find location, and has one other key attribute: the most symmetrical and memorable shape.

There are a few other planetary nebulas which look like a well-formed ring, but they are very much fainter. M57 is an astonishing sight in 4-inch and larger telescopes. When a novice searches for it using low magnification, he or she may have trouble identifying it because at such low power the Ring is smaller than the observer may think. It also requires a little bit more aperture and/or magnification to make it turn from a blurry spot to a glowing little disk to a glowing large disk with a big hole inside it. At 150× or 200× on a 6-inch telescope, M57 has become transformed into an eerie, seemingly phosphorescent cosmic smoke-ring. The band is thick, like that of a man's plain gold wedding ring, in fact, even thicker, with the ring tilted enough to plainly show the hole.

The Ring Nebula bears very high magnification well (its surface brightness is pretty high). With 8-inch, 10-inch, and larger telescopes, some detail in the nebula can be seen (for instance, it's pretty easy to notice that the ends are less bright, also that one is dimmer and projects farther—is less rounded—than the other). A fairly conspicuous star shines beyond one end of the 9th-magnitude, 80″ by 60″ nebula—this is easy in even small telescopes. More difficult is a star embedded within the opposite end. And a major challenge even for veteran observers with large telescopes is the famous "central star" of the nebula. Its brightness apparently varies between about 14.5 and 16. Walter Scott Houston proved that this hot star, which may have first puffed out the nebula only about twenty thousand years ago (the ring is still growing), is easier to see with eyes that are especially blue-sensitive (like those of most young people and especially like one of Houston's eyes in old age after its UV-yellowed lens was removed in a cataract operation and replaced with an artificial, more transparent one). What makes the central star hard to see even in very large telescopes (if they are not used at high enough power) is the glow within the hole of the Ring. Under good conditions, this glow can be noticed with just an 8-inch telescope. And with a slightly larger telescope, some of us are able to detect faint hints of up to three colors in the Ring.

Telescopic observers eager to see the Ring Nebula often overlook the naked-eye wonder of the star nearest to it in the main pattern of Lyra—Beta Lyrae. Beta is an eclipsing binary with two minima of brightness. The normal, maximum brightness is 3.4, just a tiny bit dimmer than Gamma Lyrae. But every thirty days the brightness of Beta falls (mostly over the course of a single day) to a minimum of 4.1, making it very noticeably dimmer than Gamma—until Beta brightens again rather rapidly. Halfway between two of these deep minima is a lesser one, at which Beta fades to 3.8. Beta Lyrae is the prototype of a whole class of variable stars in which two stars are so close they are pulled into ellipsoidal forms that nearly touch. A vast streamer of gas must flow from one of the component stars of Beta Lyrae to the other.

ALTAIR AND AQUILA

Although Vega is the brightest star of the Summer Triangle, the fifth-brightest star in all the heavens, and the third brightest visible from 40° N latitude, another member star in the Summer Triangle is also prominent. Altair is the eighth brightest star visible from 40° N. It marks the head of Altair the Eagle and is flanked closely by magnitude 2.7 Gamma Aquilae (Tarazed) and 3.7 Beta Aquilae (Alshain). The three stars form a short, nearly straight line, with Gamma and Beta marking the shoulders of the Eagle. The wings extend to the northwest (to Zeta Aquilae) and southeast (to Theta Aquilae).

Altair is the third closest of the 1st-magnitude stars, lying at a distance of 16.8 light-years from Earth. It shines with about ten times the Sun's luminosity. Some observers notice a slight hint of yellow in its otherwise whitish light. Astronomers have determined from study of Altair's spectrum that the star must be rotating at a tremendous rate. Although Altair is somewhat larger than our own Sun, which takes nearly a month to rotate (the rate varies at different latitudes of the Sun), Altair completes one rotation in only about six and one-half hours. This is believed to distort the star into a sort of eggshape, its equatorial diameter being almost twice that of the polar diameter.

Altair is involved with Vega in the greatest of all star myths of the Far East—the story of the Cowherd and the Spinning Damsel. The Cowherd is Altair (or the compact line of Alshain, Altair, and Tarazed in Aquila), and he is in some versions of this legend a Prince. The Spinning Damsel (spinning in the sense of weaving, not turning round like the sun Altair!) is Vega, or the little triangle of Vega and Epsilon Lyrae and Zeta Lyrae, or the whole constellation Lyra. Perhaps it is best to consider Vega alone the damsel, so that the quaint little shape of Lyra can be the loom on which she weaves.

This story, mentioned in Chinese writings more than twenty-five hundred years ago (and probably somewhat older even than that), has later versions in Japan and Korea. But the basic plot remains the same. The Cowherd or Prince and the Weaving Girl or Spinning Damsel are lovers in the sky-world who become so enamored of each other that they forget their heavenly duties. This displeases the Celestial Emperor who therefore separates them from each other on either bank of the great impassable Celestial River—and indeed Vega and Altair are on opposite sides of the bright band of that river, the summer Milky Way. But the lovers are given a single night each year to be together. It is "the seventh night of the seventh moon"—that is to say, of the seventh month, which in the Chinese calendar corresponds more closely to August than July. On that night all the magpies in the world gather and form a bridge across the Milky Way that the lovers can cross to be together. But when dawn nears, they must part again and wait until next year.

Aquila is somewhat lacking in deep-sky objects but there are some pretty doubles, one nice star cluster (NGC 6709 with a magnitude of 6.7), and one very bright variable star. The variable star is Eta Aquilae, not too far northwest of the southern wingtip marked by Theta Aquilae. Eta Aquilae is a Cepheid—a kind of star whose variations in brightness are very regular and are associated with actual pulsations of the star. Eta's brightness ranges from 3.5 to 4.4 in a period of just over seven days.

SCORPIUS AND SOUTHERN OPHIUCHUS

We have visited high Lyra and its neighbor Aquila, which is about halfway to the zenith in the southeast sky by the time dusk ends in mid-July. Now we travel to the south sky and much closer to the horizon. Looking low—on some nights through thick haze—cannot be avoided if you live at midnorthern latitudes and wish to see mighty Scorpius the Scorpion.

Even when Scorpius is at its highest and the night is clear, viewers at 40° N latitude see the southernmost part of its main pattern dimmed by more than a magnitude by atmospheric absorption. Viewers at 50° N latitude don't see this part of Scorpius at all, for it never comes above their horizon. Of the very bright constellations in the heavens, Scorpius is the only one whose main pattern is borderline in this fashion for midnorthern latitudes. It's perhaps appropriate that a constellation that is supposed to be a scorpion is seen by most of us to be scuttling along our horizon or slightly higher treeline or city skyline. In contrast, the constellation birds of summer, Aquila the Eagle and especially Cygnus the Swan, fly high.

Fortunately, some of Scorpius is much farther north, enabling fairly good views from 40° N and only slightly dimmed ones from 30° N. Scorpius helps us by having so many 2nd- and 3rd-magnitude stars. Its main pattern can be at least dimly glimpsed even through considerable haze and light pollution. And on a clear night the Scorpion is a truly striking figure above the south horizon.

The pattern of Scorpius is a gradual curve followed by a more radical one that almost curves back on itself. This second curve is the tail of the Scorpion curled upward to strike. Another way of describing the shape of Scorpius as it appears when culminating in the south is to compare it to a letter S which has fallen partway forward (that is, to the right). Under good conditions, when you look at a sky map and then the sky it is hard not to immediately recognize the constellation. But if conditions are poor or there is any doubt, you can get your bearings by noting the 1st-magnitude orange-gold star which marks the heart of the Scorpion: Antares.

Antares

Among the 1st-magnitude stars, there are only two which are red giants. One is winter's famous Betelgeuse; the other is summer's almost-as-famous Antares. There has never been much direct competition between the two because they are almost as far apart in the heavens as they could be. The very name of Antares does suggest a competition—but not between Antares and Betelgeuse. Antares means "anti-Ares"—that is, against Ares, or rival of Ares. And Ares is the Greek name for the god of war whom the Romans called Mars. It is the color of Mars which Antares rivals. The star is often brighter than the planet but when Mars is fairly near to Earth, this red planet shines many times brighter than Antares or Betelgeuse.

In the January chapter we discussed the legend of how Hera sent a giant scorpion to kill Orion and that even, after life, when both were placed among the stars, Orion would leave the sky as soon as Scorpius begins to rise. And indeed if you live at midnorthern latitudes Betelgeuse is sinking below the horizon north of west at virtually the same minute that Antares is nudging up above the southeast horizon.

Until recent years, it was always Antares which benefited from any lack of comparisons with Betelgeuse. When the two were discussed together, the statistics for Antares were always impressive but always less so than those of Betelgeuse. Antares was always second to Betelgeuse in apparent magnitude, luminosity, diameter, volume, and range of variability.

If recent distance determinations are correct, however, Antares may actually be a more luminous sun than Betelgeuse. Betelgeuse may still put out more total energy at all wavelengths (especially the infrared), though, and be cooler and larger than Antares. Also, it remains true that the apparent brightness of Betelgeuse averages about half a magnitude brighter than Antares—the former's average is 0.45, the latter's 1.06. Betelgeuse spends almost all of its time between 0.3 and 0.9, Antares between 0.9 and 1.1. The extreme values for Betelgeuse have been about –0.1 and 1.3 or 1.5, for Antares about 0.9 and 1.8. Thus Antares usually appears essentially as bright as Spica (as long as they are at similar altitudes—as is the case for observers around 40° N when Antares is at its highest) but ranks just behind that star in precise magnitude. Antares is always at least slightly dimmer than Altair and has rarely been as bright as Betelgeuse. However, the variations of Antares have been much more poorly studied than those of Betelgeuse. It's easy to understand why: most observers rating its brightness have to contend with the dimming effects of its always low altitude (not a problem for Betelgeuse) and the fact that comparison stars for it are few and far from it in the sky (moreso than for Betelgeuse). Furthermore, the main periods of these two stars are so long—about 5.7 years for Betelgeuse and about 4.7 for Antares—that it may discourage would-be brightness estimaters (it shouldn't discourage, however: there are shorter-term variations in the light of these stars and if we plan on enjoying the heavens for years why not keep track of these majestically long-period variations along the way?).

Antares may still come in second to Betelgeuse in most statistics but is impressive in its own right. For instance, Antares is so huge it would probably fill our solar system out to well beyond the orbit of Mars if we could put it in place of our Sun. Furthermore, Antares has one fascinating attribute which Betelgeuse doesn't: an observable companion star. And what a companion it is, for to many expert observers it appears vivid green.

Antares B is supposed to have a magnitude of 5.4, but it lies only 3″ from the blaze of mighty Antares. Many of today's amateur astronomers, obsessed with seeking dim galaxies and other "faint fuzzies," don't realize what a thrill it is to see this emerald in the glare of a topaz. Having a large-aperture telescope is not the key—the companion of Antares can sometimes be seen in even 4-inch telescopes— but rather to find a night of very steady "seeing." You may succeed at first in detecting only a verdant gleam flickering within one edge of the main star's image, but eventually if your optics are clean and well-collimated you should on some still summer night succeed in viewing the companion as separate.

There is controversy over whether the companion is really green or only appears so by contrast to the brighter star. There have, however, been at least a few observations in which the companion was seen as green even when it emerged from behind the dark edge of the occulting Moon alone briefly before Antares came out. By the way, the companion really is physically related to Antares, the two being estimated as having a nine-hundred-year-long orbit around each other.

The Front Half of Scorpius

Like Altair, Antares has a moderately bright naked-eye star rather close to either side of it. In this case, however, the flanking stars are about equally bright—one of them 2.8, the other 2.9. These two stars, Sigma Scorpii and Tau Scorpii, are sometimes called the Praecordia—"the outworks of the heart," the heart of course being Antares. Not far from Antares and the Praecordia are two of the sky's most interesting globular clusters—in fact, one is located just 1 1/2° west of Antares and even closer to the southeast of Sigma Scorpii.

This globular that is so close to the Scorpion's heart is M4. Magnitude ratings of it vary from about 6.0 to 5.4, but the cluster is difficult to glimpse with the naked eye because of the proximity of Antares. Binoculars, however, show it as a remarkable puff of light. M4 is one of the largest globulars in apparent size because it is possibly the closest of all to us, less than 7,000 light-years distant. (Its only rival in closeness is NGC 6397 in the constellation Ara, a much more southerly object.) Compare this distance to that of M3 (32,000 light-years), M5 (25,000 light-years), and M13 (23,400 light-years). As you might guess, the proximity of M4 helps make its individual stars easier to resolve into view than with most other globular clusters. A 6-inch (or even smaller) telescope can begin to show many of these stars. A peculiar and striking feature of M4 is a "bar" of stars across its center. With large amateur and professional telescopes, M4 becomes one of the most spectacular of all deep-sky objects, glittering with countless stars.

A less impressive globular cluster in Scorpius is M80. This magnitude 7.3 object is about four times farther than M4, so it is hard to resolve many stars in it without a rather large amateur telescope. M80 lies almost midway between Antares and the northernmost of the bright stars in the Scorpion's head—Beta Scorpii, also known as Graffias.

The head of Scorpius consists of a nearly north-south line—or, really, a gentle curve—of three brighter stars that are accompanied by a few considerably fainter ones. The line or curve is composed of Beta Scorpii (magnitude 2.4), Delta Scorpii (former magnitude 2.3), and Pi Scorpii (magnitude 2.9). Near Beta are Nu Scorpii (a tight, difficult double-double!) and Omega¹ Scorpii, both magnitude 3.9.

Notice that I said that 2.3 is the "former" magnitude of Delta Scorpii. What I meant is something amazing. As far back as we know, this star, which is also called Dschubba, shined at this brightness of 2.3. But all that changed a few years ago when Delta Scorpii suddenly started brightening and showing spectral features similar to that of another peculiar variable star, Gamma Cassiopeiae. From the summer of 2000 through the summer of 2003 when this chapter is being finalized, Delta Scorpii was usually as bright as about 1.8 to 1.6—the latter figure is equal to the magnitude of the normally second brightest star in Scorpius. Might Delta get a bit brighter and enter the ranks of the 1st-magnitude stars, or even challenge Antares? Experts think instead that Delta Scorpii may grow considerably dimmer than it originally was—maybe as dim as 3.5 or so—in 2004 or 2005 and after that perhaps only slowly brighten. But no one knows for sure. Why don't you check out Delta Scorpii tonight, comparing it first to Beta?

Before Delta's antics, Beta Scorpii was always considered the most interesting star in the head of Scorpius. It is a bright and easily split double for all telescopes. The stars are magnitude 2.6 and 4.9, 14″ apart, and most observers see them both as blue-white (they are both stars of spectral type B), though there are some interesting other appraisals of the hues. Beta Scorpii is also notable for lying very close to the ecliptic and thus being a frequent target for passing Moon and planets. As a matter of fact, both members of the Beta Scorpii system were covered over by Jupiter for viewers in parts of the Eastern Hemisphere in a very rare occultation back in 1973.

About 8 1/2° north of Beta Scorpii is a fascinating but difficult multiple star system, Xi Scorpii. The A and B stars, magnitude 4.8 and 5.1 golden F5 stars, were at their maximum separation of 1.2″ (splittable with a 4-inch telescope) in 1976 but at their minimum separation of 0.2″ (not splittable with any amateur telescope) in 1997. At the start of 2000 they were 0.4″ apart. At the start of 2005 they will be 0.8″, at the start of 2010 will be 1.0″, and at the begining of 2015 and of 2020 will be 1.2″ apart. About 8″ from this tight pair is a third star, shining at magnitude 7.3. Finally, 280″ to the south is the double star Struve 1999, a 7.4 and a 8.1 star—both deep yellow—12″ apart. And it turns out that Struve 1999 actually is a physical part of the Xi Scorpii system along with the other three stars—a marvelous true quintuple star system 100 light-years from Earth.

A Detour to Southern Ophiuchus

Since ancient times, Scorpius has been a member of the zodiac. But in truth only a small section of the ecliptic passes through the modern boundaries of the Scorpion. The Sun spends many more days passing through southern Ophiuchus, whose border comes within a few degrees of Antares. Indeed, we cannot properly discuss the vicinity of Antares in the heavens without crossing over into Ophiuchus. Most of the Serpent Bearer we studied in June. But let's look at the very rich southernmost strip of the constellation here.

Only a few degrees north-northwest of Antares, the star Rho Ophiuchi is shown in the best long-exposure photographs to be a wonderland of bright and dark nebulosity. But any small telescope can split Rho into a 5.3 and 6.0 pair 3″ apart and show a somewhat fainter star on either side of the pair (a 7th-magnitude star 156″ and 8th-magnitude star 152″ from the pair) to present the observer with an overall grouping of four stars.

An equally amazing region for both photographs and in this case also the eye lies in southern Ophiuchus about an hour of right ascension east of Rho Ophiuchi and Antares. The magnitude 3.3 star which guides us to this region is Theta Ophiuchi. A few degrees southwest of Theta is a superb triple star, 36 Ophiuchi. It consists of a 5.1 and 5.1 pair 5″ apart traveling through space with a 6.7 star that is 730″ away. All three of these stars are orange-yellow. About 3° west of 36 Ophiuchi is the globular cluster M19, a magnitude 7.1 object 4° north of the slightly brighter globular M62 (also in Ophiuchus but virtually right on the Ophiuchus-Scorpius border). As if this region of southeastern Ophiuchus were not already rich enough there is one final huge and spooky celestial object here: the Pipe Nebula.

This dark nebula is not just a sight on photographs. It can be noticed with the unaided eye in very dark and clear skies, or traced with binoculars, or its edge noticed in telescopes at low power. The Pipe Nebula is about 7° long, including a straight 5° long "stem." 36 Ophiuchi is just north of a point about 1 1/2° along the stem, working from the "mouth" end. The opposite end is the bowl, which measures about 2° wide (east-west) and 3° tall (north-south) and is centered about 2 1/2° east of Theta Ophiuchi.

The Back Half of Scorpius

South of these wonders of southeastern Ophiuchus lies the back half of Scorpius. Its stars are first southeast of Antares and the Praecordia, and run in a line almost straight south to Zeta^2 Scorpii. Then Zeta^2, Eta, and Theta Scorpii form an almost straight west-to-east line at about –43° declination. Finally a hook of stars extends back north where it is tipped with the sting of Scorpius: the side-by-side close naked-eye pairing of magnitude 2.7 Upsilon Scorpii (Lesath) and magnitude 1.6 Lambda Scorpii (Shaula). They are separated by 35′. Shaula is the 25th brightest star in the heavens. Shaula and Lesath together are known as "the Cat's Eyes." Even though one of the "eyes" is a lot brighter than the other, they are a truly impressive sight.

Within a binoculars' field north and northeast of the tail of Scorpius shine two of the most magnificent open clusters in all the heavens. These two were noticed with the naked eye in ancient times and are recorded in the catalog of Ptolemy. Today we know them as M6 and M7.

M6 is 5° north of the Scorpion's tail and glows at a total magnitude of 4.2 in a space about 20′ across. M7, only 3 1/2° to the southeast of M6, is 80′ across with a total magnitude of 3.3 or, according to Steve O'Meara's estimate, 2.8. O'Meara feels that M7 is the single brightest spot in the naked-eye visual Milky Way. It is a sprawling, loose cluster, spread out too much in telescopes, best seen with binoculars. In contrast, M6 calls for a telescope to see in full glory. And glory it is—fifty cluster members are brighter than magnitude 10.5 and one of them brighter than 6.5. The brightest stars of M6 form a stylistic butterfly pattern, leading to its nickname "the Butterfly Cluster." M7 has 5 stars brighter than 6.5 and 20 brighter than 9.0 (compared to 17 brighter than 9.0 in M6). Interestingly, although M7 has a greater apparent brightness and is much larger, it is, at 780 light-years from Earth, twice as close to us as M6 is. So the two majestic clusters are really similar in size and brightness in space.

The Cat's Eyes and the M6-M7 pair should not make us overlook the less well known region in the north-to-south line of Scorpius stars about one hour of right ascension west of them, the continuation of the Scorpion south from Antares and the Praecordia. This continuation, after Epsilon Scorpii, runs from Mu^1/Mu^2 Scorpii out to Lesath—a pattern which in Polynesia was known as the Fishhook of Maui. Maui was a god who used the fish-hook to lift some of the Pacific islands up out of the ocean's deep. Part of this Fishhook is the 4 1/2° region stretching from Mu^1/Mu^2 straight south to $Zeta^1$/$Zeta^2$. And it is one of the most marvelous places in all the heavens.

First, consider the two wide double stars. Mu^1 and Mu^2 were known in Polynesian legend as "the Inseparables," two children fleeing their evil parents. Mu^1 shines at 2.9 to 3.2 (it is variable), Mu^2 at 3.6 and the two blue-white jewels are 5.8' apart. Even more interesting is the other pair, $Zeta^1$ and $Zeta^2$ Scorpii. Those two stars are 6.8' apart and shine at magnitudes 4.8 and 3.6 ($Zeta^2$ is the star of greater apparent brightness). The stars are of spectral types B and K, respectively, and so offer a beautiful blue-white and orange contrast in small telescopes. What is amazing is that $Zeta^2$ is 150 light-years from Earth, but $Zeta^1$ is probably about 5,800 light-years from Earth—and therefore one of the most luminous stars we can see.

But it is the 1/2° long complex of clusters which extends north from the Zeta pair which is most awesome. Steve O'Meara has said it looks more like a bright comet (with a head and a bit of dust tail) to the naked eye than any other deep-sky object in the heavens.[7] The most magnificent part of the complex is the star cluster of which $Zeta^1$ is thought to be an outlying member. That star cluster remains possibly the most spectacular deep-sky object that isn't on many observing lists and still has not been seen by many amateur astronomers. Its name is NGC 6231 and its total magnitude, concentrated into just 15', is 2.6. This brightness comes from about 120 stars, including a central core area of nine stars brighter than 6.6 and ten other stars brighter than 8.6. The core has been said to look like a "mini-Pleiades." But if NGC 6231 were as close to us as the Pleiades, what would it look like? Robert Burnham Jr. points out that it would appear to be about the same size as the Pleiades, but its stars would shine about fifty times brighter than those of the Pleaides—its brightest stars would rival Sirius in our sky.

The rest of the complex of beauty just north of Zeta^1/Zeta^2 Scorpii is dominated by the clusters Collinder 316 and Harvard 12 (the latter appears as H12 on many old star atlases, but is also known as Trumpler 24). Elongated Harvard 12, which includes three 6th-magnitude stars, is the roughly 1°-long "tail" of O'Meara's "comet." Spreading away from it is the even larger (almost 2° wide) Collinder 316, a collection of stars of magnitude 6 to 9.

Now it turns out that these several clusters, and Zeta^1 Scorpii, are actually parts of a mighty O-B star association. This "I Scorpii Association" lies at a distance of 5,000 to 6,000 light-years and is therefore a bright section of a spiral arm of the Milky Way other than our own—the next arm inward toward the galactic center. Remarkably, when we look at the I Scorpii Association we are looking at it right through another, vastly more spread-out O-B association. This other O-B association is the closest of all at just 500-600 light-years, one-tenth the distance of the I Scorpii Association. It is called the Scorpius-Centaurus Association and it includes Antares; Beta, Nu, Delta, and Sigma Scorpii; Theta Ophiuchi; Alpha, Gamma, and Lambda Lupi; Epsilon, Delta, Mu, and Eta Centauri; and Beta Crucis (second brightest star of the Southern Cross). The Scorpius-Centaurus Association is believed to be only about 20 million years old, so the only one of its stars which has yet turned red giant is Antares. This association actually has stars about 90° apart in our sky and is second only to the Orion Association in apparent brightness of its members—how amazing the Orion Association stars are to outshine even these when they (the Orion stars) are two to three times farther away. The two bright, near associations in our own spiral arm—the Scorpius-Centaurus Assocation inward toward galactic center, the Orion Association outward away from galactic center—were probably created about the same time by a starburst-causing density wave passing through this section of the galaxy.

THE DELTA AQUARIDS AND THE JULY OBSERVING EXPERIENCE

How does the observing experience of a hazy, humid July night differ from that of a hazy, humid June night? Well, one way it differs is meteors. If you average meteor rates over periods of two weeks, you find that May and early June offer the year's fewest meteors but that the rate rises to more than three times greater during July. It gets even higher for the first half of August due to the addition of the peak nights of the Perseid meteors. But the Perseids are just getting started in late July and what contributes to most of the rates then are the Alpha Capricornids and, especially, several varieties of the Delta Aquarid meteors.

If you add up the total number of meteors it produces during its weeks of activity, you find that the Delta Aquarid meteor shower (or showers) is the greatest of all showers in terms of abundance. It is usually strongest in the final week of July, when one might see, in excellent conditions after midnight, one or two dozen of these often yellow meteors per hour (compare this with fifty, sixty, or more per hour at the more intense but briefer peak of the Perseids in August). The Delta Aquarid radiant is highest in the south around 2 A.M. In the wonderful shooting-star shootout of midsummer, the slower Delta Aquarids from the south will eventually lose out to the faster Perseids from the north. But in July there are more Delta Aquarids. And enough meteors to have a few bright enough to be seen even on some of the haziest nights.

Of course, there's no denying that crystal-clear skies are what astronomers usually long for. The big saviors of July nights for most kinds of stellar and deep-sky observing are the cold front passages which are all too infrequent at this time of year. When they do go through, however, there are few thrills greater than rushing outside— hopefully to a moonfree sky—and seeing the summer heavens ablaze with countless individual stars in the evening and the awesome glow of the Milky Way at its highest later in the night.

Chapter 11

August

[Constellations covered: Cygnus, Scutum, Serpens Cauda, Sagittarius]

Our road as stargazers is certain on a clear moonless August evening far from city lights. The path we should follow is marked by an exquisitely detailed band of soft but here-and-there strong glow which runs most prominently from high in the northeast to low in the south-southwest. This is the sight that cultures in human history have called variously the river of stars, the path of souls, the backbone of the night, and the spilled milk of the Roman queen goddess Juno—the Via Lactea or Milky Way.

As explained in the first section of this book, the Milky Way is both the band of light known from earliest human times and the vast spiral galaxy of several hundred billion stars which modern science has revealed that we live in. Our Sun and solar system lie within the equatorial disk of the pinwheel in which most of the stars are concentrated. And when we look in the direction of this disk we see it as a band of glow caused by the combined light of nearly countless stars too distant to be detected individually with the naked eye. The reason the band is brightest in the summer sky is that this is the direction toward the center of the galaxy and we see a greater wealth of the distant stellar multitudes than in winter, when we are looking outward from the center of our galaxy.

Suppose that your August evening is very hazy, or a fairly bright

Moon is lighting it up, or you are spending it in or near a sizable city where light pollution creates an endless glare in the sky. You will not see the Milky Way under those conditions. But you would still be able to locate the bright stars of the major constellations which lie along the Milky Way. Cygnus the Swan is the high end and Sagittarius the Archer the low end of the bright section of the Summer Milky Way. Both constellations have bright stars and striking naked-eye star patterns but they also have many additional wonders for binoculars and telescopes. In fact, no constellations anywhere in the heavens can match these two for numbers of nebulas and numbers of stars in a binocular or telescopic field of view. Few can equal them for star clusters (Sagittarius) and double stars (Cygnus) either.

In this chapter, we will spend the bulk of our time on the sights of Cygnus and Sagittarius. We won't study Aquila (we considered it along with Lyra last month). We won't study the tiny but fascinating constellations Sagitta, Vulpecula, and Delphinus, which lie in and bordering on the Summer Triangle (their coverage is in the following chapter of this book). But we will study Scutum and Serpens Cauda. These constellations on the Milky Way band are much less conspicuous than Cygnus and Sagittarius but nevertheless contain a few deep-sky treasures which are more spectacular than almost any of their kind in the heavens.

And what would be an August night without a look at the sky's most familiar and best-loved meteor shower, the Perseids?

CYGNUS

Cygnus the Swan is the standout constellation on high after Vega and little Lyra pass the zenith. Throughout the summer we watched the Swan's steady ascent up the northeastern and eastern sky. We noted Cygnus and the placement of its 1st-magnitude star, Deneb, in the larger, unofficial star pattern called the Summer Triangle. On August and September evenings, however, Cygnus soars overhead for much of the world's population—everyone at midnorthern latitudes. These late-summer evenings are therefore superb ones for studying the many wonders of the heavenly Swan.

The main stars of the Cygnus pattern can be detected with the naked eye even when the sky is fairly bright from moonlight or light pollution. The shape really does resemble the outline of a long-necked graceful bird in flight. Alternatively, this main pattern of Cygnus is "the Northern Cross," with the body and neck of the bird the upright of a cross and the wings the tranverse bar. However, it is not until Cygnus reaches the northwest horizon that the Northern Cross itself seems to stand straight up with respect to the horizon. This sight, as Cygnus sets, can be seen in the early evening in December but on nights of late summer it is not visible until dawn is near.

When Cygnus is high on August evenings, it can be more naturally regarded as a swan. Even though Deneb is the brightest star in Cygnus, it marks not the head but the tail of the bird. Indeed, the name Deneb is derived from the Arabic for "tail." The beak or eye of Cygnus is represented by the star Albireo, which shines at only third magnitude but which is, as we shall see presently, a star of great interest for reasons other than brightness. In any case, the Swan pictured by the stars of Cygnus is heading southward or, to be more precise, south-southwest late on August evenings. Many swans do migrate south at this time of year, and many birds follow the landmark of a coastline or river when they migrate. In the case of Cygnus, the river is the heavenly river of the Milky Way, for the Swan seems to be headed straight down the glowing path of the Milky Way band. The wings of Cygnus stretch roughly from one "bank" of the Milky Way river to the other. And around the long neck of the Swan glows a bright cloud, seemingly of mist, but really of countless stars.

This is the Cygnus Star Cloud, one of the brightest of the Milky Way band. And, for people at midnorthern latitudes, it is the most prominent star cloud when summer haze takes its toll on the Sagitarius and Scutum Star Clouds lower in the sky. The Cygnus Star Cloud can often be glimpsed dimly with the naked eye even from a typical suburban location or medium-size city. On a clear, moonless country night, it glows with wonderful intensity and structure.

Deneb and Albireo

The two ends of Cygnus the Swan are fascinating. For naked-eye observers, the star at the northern end, which is much brighter, is the one which grabs attention.

That star is Deneb. It is the dimmest of the three members of the

Summer Triangle and, at magnitude 1.25, one of the dimmest of the 1st-magnitude rank. Appearances can be deceiving, however. In reality, Deneb is roughly twice as distant as any of the other 1st-magnitude stars and is therefore the most luminous of them all. If Deneb is about 1,700 light-years away (it may be farther), then it is approximately a thousand times more distant than Altair, the star in the Summer Triangle which outshines it by about a half-magnitude in our sky. The light Deneb releases to the universe in the course of one August night is about as much light as our Sun will produce in the entire twenty-first century!

The jury is still out on whether mighty Deneb is the illuminating source for a region of gas which glows 3° east of it—NGC 7000, the North America Nebula. The nebula earns its name from the resemblance of its shape to that of the continent. You can notice the nebula with the naked eye in a dark country sky but it looks just like a bright patch of Milky Way and you won't be able to notice the details of its outline without magnification. The North America Nebula is a beautiful sight in good binoculars and wide-field telescopes, though even under the best of conditions a little elusive for beginners who haven't trained their eyes to see the sights of the sky.

The star marking the opposite end of the Northern Cross or Swan pattern seems to the naked eye a modest point of light. Beta Cygni, better known as Albireo, shines at a magnitude of 2.9. A look at Albireo with a little magnification begins to reveal that it is no ordinary star, however. Use steadily held or mounted binoculars of 7× or higher power and you should be able to see Albireo's point of light split in two. The higher magnification and greater light-gathering power of a small to medium-size telescope is needed to give a comfortable view of what many observers consider to be the most beautiful colored double star in all the heavens.

Albireo is a magnitude 3.1 gold star shining 34″ from a magnitude 5.1 blue star. When savoring this celestial showpiece, there is telescopic advice to bear in mind. First of all, low magnification is all that is needed on your telescope for such a wide double. Double stars are usually most beautiful when the apparent separation between them is small. Second, the greater light-gathering power of large amateur telescopes actually tends to wash out the color of such relatively bright stars as the components of Albireo. It's pleasing to view Albireo with an 8-inch telescope because it will display smaller, more precise star images. But the most intense colors in a double star this bright might be best seen with an aperture of 4 or 6 inches. A special trick to try also is making the star images *less* sharp—that is, slightly defocusing the stars. This sometimes seems to give a greater strength to the colors perceived. Of course, what an observer thinks is the finest view of a splendid color-contrast double like Albireo depends also on the physiology of the particular observer's eye and, most important of all, on his or her psychology. The beauty of a sight like this certainly has a subjective element to it for observers.

Until a few years ago, the only major interest in Albireo was in its appearance. But then the positional measurements of stars made by the Hipparcos satellite in the 1990s enabled astronomers to determine much more accurately the distances to stars and the motions of stars. And this in turn made it possible to calculate which stars would be the brightest in Earth's skies for up to several million years into the past and future. Like Gamma Draconis (see the June chapter), Albireo in a few million years will come much closer and become the brightest star in Earth's skies. Unless the angle of perspective on Albireo is then unfavorable, observers on Earth should then be able to see with the naked eye (if it is anything like the current human naked eye!) an amazing sight. They will see the gold component of Albireo as a zero-magnitude star with the blue component as a 2nd-magnitude light huddled next to it.

Other Double Stars of Cygnus

There are many fine double stars in Cygnus. As a matter of fact, there is a double in Cygnus whose colors and magnitudes are somewhat similar to Albireo's but which is part of a more complex arrangement with two other colorful stars!

The system I refer to as a "double" is the one called Omicron[1] Cygni. The confusing thing is that when you look with the naked eye you see a pair of magnitude 3.4 and 4.0 stars, Omicron[1] and Omicron[2], about a degree apart and several degrees northwest of Deneb—but these are not the stars you see as a pair when you look through the eyepiece of most telescopes. For at telescopic magnification, Omicron[2] may be out of your field of view (or at least appear far removed from Omicron[1]) and what you may first regard as a pair are the brightest components of Omicron[1] Cygni: a 3.8-magnitude gold star that is

338″ from a delicately green 4.8-magnitude star. Then you quickly notice that 107″ from the gold star is a 7.7 magnitude star of a different and lovely hue. Some people call this star blue, but in some instruments it seems to me a stunning purple. I discovered the beauty of the Omicron[1] trio for myself about a quarter-century ago (only in the past few years has the system started becoming publicized and known to more observers). The next day, I was walking in the June sunshine when suddenly I saw a variety of small flowers (perhaps some kind of violet) a few feet away from me in a roadside field. The deep purple of the flowers suddenly brought to mind the color of the 7.7-magnitude component of the Omicron[1] trio with an immediacy which took my breath away. It was almost as if the color literally leaped from the flowers to my eye, filling me with recognition and wonder.

But I haven't finished describing the full scene you can behold when you look at the Omicron[1] system! For if you have a wide enough field of view (in binoculars or rich-field telescope) or with a narrower field of view can move the telescope back and forth, you can also compare the colors of the Omicron[1] trio to that of magnitude 4.0 Omicron[2] Cygni—a vividly orange star.

Further confusion about the complex Omicron[1] and Omicron[2] systems has occurred for several reasons. For one thing, many books used to (and perhaps some still do) refer to Omicron[1] and [2], respectively, only as 31 and 32 Cygni. It's also possible to confuse Omicron[1] and [2] Cygni with nearby Omega[1] and [2] Cygni, partly no doubt on the basis of the similarity of the names. But, amazingly, the Omega[1]/Omega[2] Cygni complex does resemble Omicron[1]/Omicron[2], though the former is less bright (5th-magnitude stars) and tighter. Omega[1] is 20′ from Omega[2]. Omega[2] is itself a double consisting of a deep orange magnitude-5.4 star and a white (or blue-white) magnitude-6.6 star a full 257″ apart.

The Omicron Cygni pair is only about 3° from the Omega Cygni pair. The two form a triangle with Deneb in which Omicron is 5° and Omega is 4° Deneb.

Still more could be said about the wondrous Omicron[1] and [2] Cygni and their vicinity. For instance, Burnham notes in his classic celestial handbook[8] that 45″ to the east and slightly north of Omicron[2] is the very red variable star U Cygni (magnitude 5.9 to 12.1 over a period of 462.4 days) whose "ruddy glow" makes "a striking contrast with the 7.8 mag bluish star 65″ distant." Much more could be said about the distances and physical nature of the stars making up the Omicron and Omega Cygni systems. The two brightest stars of the Omicron[1] trio are approximately 1,400 and 750 light-years from Earth, and Omicron[2] is about 1,000 light-years from us. Both Omicron[2] and the primary star of Omicron[1] happen to be vast K-type giants which, though of slight visual variability, are eclipsing binaries—Omicron[2] has a period of 3.12 years, Omicron[1] a very lengthy 10.42 years! And both these eclipsing binaries have an extended outer atmopshere or corona of the big stars which causes a partial eclipse of their companions on either side of the total eclipse.

To the opposite side of Deneb from Omicron and Omega Cygni (and much farther from it than those stars) is a double star both beautiful and consequential in the history of astronomy. The star is 61 Cygni. It is one of the stars of greatest proper motion—its change of position on the celestial sphere due to its movement through space is 5.22" per year. In fact, it held the title of star of greatest proper motion from the time its motion was measured by Giuseppe Piazzi in 1792—he called it the "Flying Star"—until an even faster star was identified in 1842. Even more important than its proper motion is the fact that 61 Cygni was the subject of the first reasonably successful attempt to estimate the distance to a star by its parallax (the change in its apparent position caused by us viewing it from opposite ends of Earth's orbit during the year). That success was achieved in 1838 by F. W. Bessel (Bessel was also the most meticulous observer and detailed sketcher of Halley's comet in 1835–36). Bessel calculated that 61 Cygni is 10.3 light-years from Earth. Today we know the correct figure is a bit farther, at 11.4 light-years—essentially the same distance as Procyon. Only three star systems readily visible to the naked eye— Alpha Centauri, Sirius, and Epsilon Eridani—are closer to Earth.

So 61 Cygni is a close, fast, historically important star system. But what is it like visually as a double star? Beautiful. Both stars are orange, with one shining at magnitude 5.2 and the other at 6.0. Their separation is 30", a distance the pair travels across the sky together in just six years of proper motion.

Cygnus has many other fine double stars. Tight but bright ones include Delta Cygni and Mu Cygni. Much wider and easier for beginners is 16 Cygni, 6.0 and 6.1 G-type stars that are 39" apart. Double-star expert James Mullaney calls 16 Cygni a "perfectly-matched, roomy golden pair"[9] and notes that they are also a good guide for finding a fascinating nebula—the famous "Blinking Planetary."

Nebulas and Clusters of Cygnus

The "Blinking Planetary" is the name Mullaney himself applied to this object (in 1963), and he was the person who popularized the nebula. Its official designation is NGC 6826. The nebula itself has a total magnitude of about 8.9 with its central star a bright 10.4. To see the planetary "blink" requires at least a 2.4-inch or 3-inch telescope (and excellent conditions) but it gets better and better as aperture is increased further. What is the blink? When you stare directly at the central star, the nebula seems to disappear. Then when you look a bit to the side (use averted vision) the nebula suddenly pops back into view, so brightly that it overhwelms the sight of the star! There are other planetary nebulas which can produce this blinking effect but none do it so brightly or easily as NGC 6826. A line extended from Deneb through the Omicron Cygni pair can be continued far onward until they come to the wingtip stars of Cygnus which are near the Blinking Planetary.

We have already discussed the famous North America Nebula. Another famous nebula in Cygnus is actually for amateur observers a few of the brightest sections of a larger complex of nebula. Long-exposure photography reveals the entire structure to be a big 3°-wide bubble produced by a supernova, which may have flared in our skies something like 150,000 years ago. Parts of this Cygnus Loop or its (visible) entirety have been variously named the Filamentary Nebula, Lacework Nebula, Network Nebula, Cirrus Nebula, and Bridal Veil Nebula. But the most common name is the Veil Nebula.

The easiest segment of the Veil for beginners to find is the west section, NGC 6960, for it is a strand which runs right through the magnitude 4.2 star 52 Cygni. The east part of the Veil is NGC 6992-5 and it is the brightest. These most obvious parts of the Veil Nebula can be detected in surprisingly small wide-field telescopes—even in 7× binoculars—if your sky is very dark. Increase the aperture to 6 inches or 8 inches under good conditions and you may start seeing a little bit of structure in these ghostly arcs. With nebula filters, more structure becomes evident. However, it really takes rather large 16-inch or 18-inch amateur telescopes to begin showing so much filamentary structure that the strands of the Veil take on a three-dimensional appearance, in one part looking like a luminous tornado tube frozen in space. With even larger telescopic apertures, colors begin to be glimpsed. Long-exposure color photographs show vivid red, white, and blue hues in the filaments of the Veil. These glowing shreds of an ancient supernova lie about 2,500 light-years from Earth.

There are other patches of luminous nebula in Cygnus and the dark nebulas which indent the Milky Way in this constellation are fascinating. Here is where begins the Great Rift, which splits the summer Milky Way into a main channel that goes on down to the horizon and a branch which extends all the way to Ophiuchus. But don't neglect to also notice the dark indentation which seems to stretch almost all the way across the Milky Way band about 7° north of Deneb.

Cygnus is crawling with open star clusters. The only problem is that the star fields in the constellation are so rich that many of these clusters are difficult to distinguish from the background. There are regions of Cygnus which probably display more stars in binoculars than any other places in all of the heavens.

One cluster in Cygnus certainly deserves mention and attention even from beginners. Way up in the northeast corner of the constellation is M39. This cluster is about 9° east-northeast of Deneb and 3° north of magnitude 4.0 Rho Cygni. If you get your gaze into this area in a dark sky, you should see M39 as a magnitude 4.6 fuzzy patch of light. The cluster was apparently recorded in ancient times by Aristotle. M39 is best seen with the low magnification of small telescopes and binoculars, for it is a large (1/2°) and sparse grouping. It contains about thirty stars but the dozen or so brightest are of similar brightness and contain a pattern resembling that of the famous Pleiades cluster within a triangular outline.

Variables and Other Strange Stars in Cygnus

The summer Milky Way is a hotbed for novas, exploding stars. Two particularly bright nova explosions occurred in Cygnus in the twentieth century. Oddly, both flared up to a maximum magnitude of about 1.8 and both during the last week of August: the first peaked on August 24, 1920, the other on August 30, 1975. Nova Cygni 1975 brightened by almost a magnitude in six hours the night before it reached peak brilliance. These objects dramatically changed the appearance of Cygnus, though only briefly.

Cygnus is home to one of the superluminous stars which have been called "permanent novas." P Cygni was first noticed as a star of about magnitude 3.0 in the year 1600. Its magnitude has fluctuated between that value and about 6.0 in the four hundred years that have followed (though only between 4.6 and 5.6 in the past two centuries). P Cygni is believed to be about 5,000 light-years away and therefore one of the most luminous stars known. It is probably most like another shell-producing star which has flared up to great brightness, the far-south wonder Eta Carinae.

The second-brightest of all the long-period variables with huge ranges is Chi Cygni. Only Mira (Omicron Ceti) typically gets brighter. Chi Cygni usually ranges from about fourth or fifth magnitude down to dimmer than twelfth magnitude. It can become as dim as about magnitude 14.2 but on rare occasions—such as in 2002—Chi Cygni can flare up to become a 3rd-magnitude object. Its range is greater than that of Mira. Its period is considerably longer—407 days. When this somewhat ruddy star has a bright maximum, it becomes part of the neck of the Swan just 2 1/2° headward (that is, closer to Albireo) from similarly bright (magnitude 3.9) Eta Cygni.

A bit less than 1/2° east-northeast of Eta Cygni lurks an astrophysical beast of great fame—Cygnus X-1. This powerful X-ray radio source seems to be the unseen companion of a 9th-magnitude star. And that unseen companion is almost certainly a *black hole*. No star is stranger than a black hole: the ultimately collapsed remnant of a supernova. Cygnus X-1 was the first object suspected, and first object almost conclusively confirmed, to be a black hole.

SCUTUM AND SERPENS CAUDA

We regretfully leave Cygnus and head south down the Milky Way across Aquila. The neck of Cygnus points us through the stars of the northwest wing tip of Aquila to Serpens Cauda, the Serpent's Tail. And just southwest of the tail stars of Aquila is the little constellation Scutum the Shield. Scutum and Serpens Cauda can also be found by scanning south down the admirable length of the Milky Way in a dark sky. In such a sky, Serpens Cauda is a line of stars shining in front of the darkness of the Great Rift of obscuring gas and dust which splits the Milky Way band. And Scutum is dominated by the roundish, remarkably bright glow of the Scutum Star Cloud.

The Scutum Star Cloud and M11

The Scutum Star Cloud measures at least 5° by 5° and is seen about halfway up the south sky by observers at midnorthern latitudes in August evenings. On some hazy summer nights, the grand star clouds of Sagittarius can be low enough in the sky to be washed out by haze while the Scutum Star Cloud still shines bright. It excels not only by its brightness but by its concentration and form. The great observer E. E. Barnard referred to the Scutum Star Cloud as "the gem of the Milky Way."

Right in the midst of the Scutum Star Cloud and star pattern of Scutum shines a star cluster brighter than the naked-eye limit but somewhat difficult to detect without optical aid thanks to the background brightness of the star cloud. The cluster becomes easily evident in binoculars. But it is in a medium-size or larger telescope that this M11 in Scutum is shown to be possibly the richest and most splendid of all bright open clusters. It is my own personal favorite of all star clusters in the telescope.

M11 has been nicknamed the Wild Duck Cluster. The title is derived from a phrase of one of the two greatest nineteenth-century writers on deep-sky objects, Admiral William Henry Smyth. Smyth described the cluster as resembling "a flight of wild ducks." Most of the brighter stars of M11 do indeed form a sort of triangle or V-shape like that of a flock of wild ducks or geese. The leader of this flock is an 8th-magnitude yellow star near the apex of the fan.

A small telescope at low magnification shows M11 as mostly just an interesting hazy condensation, which one could mistake for a globular cluster, in a magnificent section of heavens. Increase the aperture of telescope just a bit, however—say, to 6 inches—and things start to get thrilling. A medium-size telescope begins to reveal dozens upon dozens of individual stars but leaves a charming background glow from the combined light of fainter stars. I once described this view as being "like an avalanche of stars throwing up a cloud of nebulosity." The avalanche is falling, of course, from the the apex of the bright fan of M11 and its brightest star.

Now when an observer uses a fairly large telescope—10-inch aperture or more—the minimum magnification usable is not very low and most open clusters no longer look rich because their stars are shown too widely spread. But M11 is the great exception. As Walter Scott Houston wrote, in a larger telescope M11 appears as "a carpet of sparkling suns to the very center with outlyers swarming on all sides. A good 10-inch shows hundreds of glittering star points all over the field of view."[10] Even those who have observed M11 at very high power through 30-inch and 36-inch refractors have found that the cluster still looks rich. And the statistics about M11 bear this out. Its total magnitude may be only about 5 1/2 but within about 10' to 13' of sky M11 offers approximately five hundred stars brighter than 14th-magnitude (most of which are in the 11th- to 13th-magnitude range). About four hundred of these stars are true members of the cluster. But a 1977 study found the number of stars brighter than 16.5 that are true members of M11 is about 870.

While you are enjoying M11 in its lovely setting with low power, be sure to look near it for eerie regions almost devoid of visible stars—obscuring clouds of dark nebulas which Edward Emerson Barnard discovered and cataloged. Consider also the distances of the objects you are seeing. The dark nebulas are between us and the Scutum Star Cloud. M11 lies about 5,500 light-years from Earth but it, too, is much closer to us than the Star Cloud.

What would it be like to live on a planet circling one of the stars of M11? Star cluster expert R. J. Trumpler calculated an answer which has often been quoted. According to Trumpler's estimates, an observer near the center of M11 would see several hundred 1st-magnitude stars and maybe forty or so shining from three to fifty times as bright as Sirius does when viewed from Earth! The very brightest stars in that sky would easily outshine Venus.

M16 and Other Wonders in Serpens Cauda

When we turn from Scutum to Serpens Cauda we are moving into the Great Rift in the Milky Way, so the views are much less rich in stars. About 10° due north of M11 and the modest 4th- and 5th-magnitude stars of the Scutum pattern shine a few interesting deep-sky objects in the northern part of Serpens Cauda. One of them is 4.0-magnitude Theta Serpentis, which is also known as Alya. Alya is a lovely, easy double consisting of a 4.5 and 5.4 star, a comfortable 22" apart. Both stars are of A5 spectral class. The hues of both have been very variously named by different observers: are they yellow, white, or blue-white? Take a look and see what you think.

Between Theta Serpentis and 72 Ophiuchi is a little-known but remarkable pair of big, low-power clusters which are plainly visible to the naked eye in really dark skies. The large branch of Milky Way separated from the main band by the Great Rift can, under superb conditions, be seen extending very faintly over Beta and Gamma Ophiuchi and even deeper into Ophiuchus. But the brighter part of the branch ends well northeast of that area. And marking this end from within it are two sizable spots of hazy light: the S-O Double Cluster.

S-O Double Cluster is short for Serpens-Ophiuchus Double Cluster, for one of these objects is in Serpens Cauda and the other just over the border in Ophiuchus. The two are separated by 3° and their magnitudes and sizes are very impressive, though they are loose, low-magnification objects. NGC 6633 in Ophiuchus is magnitude 4.6 and 27′ wide, and contains at least thirty stars, the brightest of which is magnitude 7.6. IC 4756 in Serpens is about magnitude 5.1 and at least 52′ wide, and includes at least eighty stars, the brightest of which is magnitude 8.7. Make certain that you don't confuse this S-O Double Cluster—just named and noticed as a wide pair in recent years—with the more famous and much more magnificent Perseus Double Cluster. The Perseus Double Cluster features a pair far closer together and tremendously richer than this one in the summer sky. Yet the S-O Double Cluster does have a charm of its own.

To visit one more deep-sky object in Serpens Cauda, we must travel from the northern reaches of Serpens Cauda down to its southern border. South of that border is the mind-bogglingly rich Sagittarius Milky Way, with more visible nebulas and star clusters than perhaps any other comparably sized region of the heavens. But this last object in southernmost Serpens is really a first of that Sagittarius region, even if it is technically not quite within the bounds of Sagittarius. The object to which I refer is M16, also called the Eagle Nebula or the Star-Queen Nebula.

M16 is really a nebula and cluster together, just visible to the naked-eye as a magnitude 6.0 spot of glow. The nebula has long had a reputation for being very elusive in small telescopes, though much depends on the darkness of the sky. Nowadays, most advanced amateur astronomers will use a nebula filter here with good results. Nevertheless, it is important to remember that this nebula has earned its dramatic names only from its appearance on long-exposure photographs. These images are among the most impressive of any object in the heavens. And the awe of them has been deepened by the most spooky of all Hubble Space Telescope images: those showing M16's intricate fingers of dark gas at whose tips new stars are being born.

SAGITTARIUS

Finally, rather low in the south as seen from midnorthern latitudes, we reach the widest, most complex, and overall brightest portion of the heavens-encircling Milky Way band. When we look at this region, we are catching a glimpse of the galactic hub and staring toward the galactic center. The center itself, an unimaginably glorious ball of many billions of stars and raging mystery (astronomers firmly believe it includes a massive black hole), is hidden from us by a few tens of thousands of light-years of absorption from clouds of interstellar dust and gas. But even the objects just one-third to one-half of the way from us to the center which we see here—in northwestern and northern Sagittarius—form a congregation of grandeur in many ways unmatched anywhere else in the heavens.

The Teapot of Sagittarius

Let's begin with the nearby stars which make the shape of Sagittarius. These stars are all that you will see with the naked eye in this part of the heavens if you are observing in heavily light-polluted conditions. Even in dark skies they are useful as guides to locating clusters and nebulas—in addition to being attractive in their own right.

For indeed the major stars of Sagittarius the Archer—which include 2nd- and 3rd-magnitude luminaries—make a quite striking pattern. In ancient times some of those stars were used to form a bow and its archer, who happens to be a centaur (half-man, half-horse). In recent decades, however, it has been noticed that the most important stars of Sagittarius form a very convincing outline of a kitchen item completely unrelated to centaurs, bowmen, and Greek mythology. The major stars of Sagittarius form the asterism known as the Teapot.

The Teapot of Sagittarius is immediately evident on a star map (if the lines connecting stars are drawn in) and also in the sky—once it is pointed out to the beginner. Its brightest stars are magnitude 1.8 Epsilon Sagittarii (Kaus Australis) and 2.0 Sigma Sagittarii (Nunki). Sagittarius is perhaps the most extreme example of a constellation

Figure 11.1. Perhaps the spookiest of all Hubble Telescope images is this one of a part of M16, showing pillars of gas at whose tips new stars are being born.

with stars whose Greek letter designations do not correspond to their order of brightness (Alpha and Beta Sagittarii shine at weak magnitudes of 4.0 and 3.9 and are far south of the stars of the Teapot).

The Teapot can be pictured as blowing a steam of nebulas and star clouds up from its spout. A similar imagining is that the Milky Way is the smoke coming up from Ara the Altar, a constellation below Scorpius which is too far south for us to consider here. But during about the first three-quarters of the twentieth century there was another asterism made from some of the stars of Sagittarius which was very popular and useful. I refer to the Milk Dipper. The bowl of this asterism was made of the four stars which form the handle of the Teapot. But the handle of the Milk Dipper ran to Lambda Sagittarii— the top of the Teapot—and then onward northwest to one additional, fainter star that is not in the Teapot. That star is Mu Sagittarii. It sits right in the center of a circle of many of the finest deep-sky objects of Sagittarius, and so is a useful guide to finding them.

The Star Clouds and Other Naked-eye Wonders

Imagine that you are out at a very dark site and viewing the entire Milky Way region of Sagitarius and eastern Scorpius. You are staring at the face (or one face) of the largest physical structure the unaided eye can really see much structure in and begin to comprehend—the vast galaxy in which we live. Even if you know nothing about what you are seeing, however, there is a grandeur in this view which stirs the imagination like no other. You may think at first that you are observing clouds of water vapor or ice crystals in Earth's atmosphere that are lit by some earthly source. But you are really seeing clouds of *stars*. The entire band of Milky Way in this part of the sky is the combined glow of literally millions of stars. And here and there we behold brighter, more concentrated patches, which we do refer to as "star clouds."

Farther north along the Milky Way we already met the Cygnus and Scutum Star Clouds. Stretching from just north to northwest of the Teapot's spout is the overall brightest of them all, the Large Sagittarius Star Cloud, an impressive big region of glow beyond which lies the awesome center of our galaxy. Almost 12° due north of the start (the east side) of the spout is what is possibly the brightest knot of the naked-eye Milky Way—the Small Sagittarius Star Cloud. This 2° by 1° patch of stars has earned its own Messier number—M24. Dramatic

Figure 11.2. A spectacular section of Milky Way, photographed by Akira Fujii.

with the naked eye, marvelous in binoculars, M24 fills a wide tele-scopic field of view with more easily, individually visible stars than perhaps any other place in the heavens. Note also the presence of dark nebulas near this brilliant region. It's hard to visually estimate the total magnitude of an object as extended as M24, and Steve O'Meara is surely right when he rejects the 4.5 figure one often reads. He puts M24 at magnitude 2.5. Still, don't forget that this is spread out over an area several times larger than the Moon's apparent size. The surface brightness of M24 is not great enough for it to be seen with the unaided eye in a light-polluted sky.

How many other exotic spots or patches of glow in Sagittarius and eastern Scorpius are visible to the naked eye in a dark sky? Many, if your conditions are excellent and you search them out carefully. But for beginners, let's mention just those which should be *easily* visible if your light pollution is minimal. In the chapter for last month, we noted that M7 is plainly visible without optical aid, and its fellow cluster, M6, to a lesser degree. But you should definitely also spot a sizable puff of glow just above the Large Sagittarius Star Cloud or about 6° almost due north of the star which marks the tip of the Teapot's spout. This puff is about 6 1/2° southwest of the considerably larger M24. Its identity? It is summer's brightest cloud of gas, M8—the Lagoon Nebula.

Nebula Row

Four of what are arguably the five most famous diffuse nebulas visible from midnorthern latitudes are located along a few-degree-wide "street" of sky less than 12° long. M42, the Great Orion Nebula, is a winter object. But in northern Sagittarius we find a straight line, running about south-southwest to north-northeast, that consists of the nebula M8, the star cloud M24, and the nebula M17. And just to the north (or a bit west of north) of M8 and M17 are the other two famed nebulas, M20 and M16, respectively. M8 is about 6° from M24, and the eye must travel another 2 1/2° from M24 to M17. M20 is perched only about 1 1/2° from M8. And M16, though it is over the border into Serpens, is only 2 1/2° from M17.

Don't get lost in this welter of numbers. Amateur astronomers come to relish the Messier numbers of their favorite deep-sky objects, but the names of these nebulas make them come alive—as do the actual sights of them.

The first case in point is M8, better known as the Lagoon Nebula. This huge cloud of gas is second only to the Great Orion Nebula in overall brightness and grandness among the nebulas visible from midnorthern latitudes. Its photographic dimensions are an amazing

80' by 60'. Its surface brightness is not as great as the Orion Nebula's, though, and most of us see it fairly low in the sky and often through summer haze. In poor sky conditions, little of the nebulosity may be detected in a small telescope, though the 4.6-magnitude star cluster NGC 6530 (probably physically involved with the nebula) is visible. Get good sky conditions and the sight is transformed into a dramatic one, especially if you have at least a medium-size telescope. The star cluster occupies the eastern, dimmer section of the nebula. The two sections of nebula are dramatically separated by a "lagoon"—really a channel—of dark cloud. The nebula is stirred to radiance by one visible and perhaps a few more hidden stars. The visible star is 9 Sagittarii. This magnitude 6.0 star is a very hot O5 sun. Near it is the brightest part of M8, a 30" section which, due to its shape, is called the Hourglass. Can you see a hint of red or pink hydrogen emission in parts of M8? To detect a lot more structural detail, use a nebula filter. M8 is, like M42, a hotbed for starbirth. Amateur astronomers with large telescopes and nebula filters have even detected some of the dark "Bok globules"—clouds which are thought to be stars in their initial stages of formation.

Remarkably, another major nebula lies only 1 1/2° north-north-west of M8, and so can be fit into the same very wide telescopic field of view with the Lagoon. This is M20, the Trifid Nebula. Just 0.7° northeast of M20 is the fine 5.9-magnitude open star cluster M21, which adds greatly to the splendor of this scene. Though smaller and less in total brightness than M8, the Trifid Nebula itself is an impressive sight in medium-size and larger telescopes. It owes its name to the dark lanes which split it into three parts. An eight-inch or larger telescope and good conditions are in order to see these plainly and to glimpse hints of red and blue in different parts of the nebula. (Don't expect to see colors or details in these nebulas with anything like the intensity they appear in photographs, of course—be patient and slowly learn to see more and more through careful inspection.)

Near the place where the dark lane in M20 forks into two lanes is the triple star HN 40, whose brightest star may be the principal source of radiation exciting the Trifid to glow. HN 40 is an obvious double in small telescopes, 7.6 and 8.7 stars 11" apart. Steve O'Meara sees the brighter star, the hot O7 sun which stirs the Trifid's glow, as mustard-colored for some reason and the 8.7 star as "a dying ember, charcoal-colored with a spark of red."[11] The third star, magnitude 10.7, can be seen in medium-size telescopes. Astronomers have established the existence of three more stars in the system, thus matching (though much more dimly) the total of six found in the Orion Nebula's multiple star Theta[1] Orionis (the Trapezium). It's interesting to note that M8 and M20 are both at roughly 5,000 light-years from Earth, almost four times farther from us than M42. And that is also the approximate distance of our next nebula, M17.

M17 is probably most often known as the Omega Nebula. But it is also called the Horseshoe Nebula, Swan Nebula, and Checkmark Nebula! All of these are attempts to describe its dramatic shape (the omega referred to is the capital form of the Greek letter omega—Ω). Other observers have compared the form of the nebula to that of the number 2. It is the bottom stroke of the 2 or body of the swan which is immediately apparent in small telescopes, for this is a surprisingly intense feature. Further inspection (or a bit more aperture) readily shows the fainter hook (the swan's neck and head) at one end of this bar of light. The best observers can in very dark skies see much more luminosity around the main shape of the nebula even with fairly small telescopes. Alternatively, larger telescopes and nebula filters can do the trick, too.

The final object in our "nebula row" is the one we discussed earlier in this chapter, for technically it lies within the bounds of Serpens Cauda not Sagittarius. M16, the Eagle Nebula or Star-Queen Nebula, is believed to lie about 9,000 light-years distant, almost twice as far from us as the Lagoon, Trifid, and Omega Nebulas.

The Clusters of Sagittarius

We've already met with a few fine clusters that lie in or near the great nebulas. But the constellation of the Archer has others which are among the sky's showpieces.

Best of these is surely the great globular cluster of Sagittarius, M22. It is located just 2 1/2° to the northeast of Lambda Sagittarii, the star which marks the top of the Teapot. Only its relatively low position in the summer sky, where haze will often take a toll on splendor, has prevented M22 from being as popular and praised as M13 high up in Hercules. Which cluster is greater? M22 is at least a little brighter (estimated magnitude: 5.2) and is bigger (as much as 33' across). More of M22's stars are resolvable because they lie just over 10,000 light-years away (less than half the distance of M13). Still, I would not wish to choose between favorites and would instead stress that both M13 and M22 have different grand appearances. Be sure to observe M22 on a really clear night when it is at its highest. If you do, you may agree with Robert Burnham's suggestion that M22 in the telescope looks like the great jewel called the Arkenstone of Thrain as described by master fantasy writer J. R. R. Tolkien in *The Hobbit*: "It was as if a globe had been filled with moonlight and hung before them in a net woven of the glint of frosty stars."[12]

One fairly bright globular cluster in Sagittarius is M55 but it is hard to find all by its lonesome in star-poor southeastern Sagittarius. A few lesser globular clusters—M54, M70, M69—decorate the bottom of the Teapot. In northern Sagittarius, however, are several open clusters of great beauty.

Perhaps the two best of these open clusters are located to either side of M24, the Small Sagittarius Star Cloud. M23 is about 5° due west of M24. It is a magnitude 5.5 cluster which includes about 150 stars in a space 30' wide. M25 is a favorite of mine. It shines about 3 1/2° east-southeast of M24. It has a total magnitude of 4.6 and is about 1/2° across. It is a fairly easy naked-eye object. What makes it really beautiful is its thirty or so brightest stars and the dramatic arcs and bunchings they assume. A true member of the cluster is the Cepheid variable star U Sagittarii, which varies from magnitude 6.3 to 7.1 in a period of 6 days, 18 hours.

THE PERSEIDS AND THE
AUGUST OBSERVING EXPERIENCE

After an evening of contemplation and study of the innumerable distant, fixed wonders which adorn the summer Milky Way, the latter part of August nights in the first half of the month offer a strikingly different entertainment: swift and startling meteors.

As the Delta Aquarid meteors which peaked in late July gradually lessen, another meteor shower starts flooding out from a region between the main patterns of Perseus and Cassiopeia. This wonderful Perseid meteor shower is often the strongest meteor shower of the year and is certainly the most famous and most often watched by the general public. After all, this is a time of summer vacations, outdoor activities, and comfortable nights. The Perseids peak on August 12 or 13 each year and if your skies are clear and dark, forty, fifty, or even sixty or more Perseids per hour may be glimpsed in the best hours after midnight. The radiant or source point of the meteors is at its highest around 4 A.M., not long before morning twilight begins at midnorthern latitudes.

Let the Perseids become a summer tradition for you and your family and friends. While you watch these swift streaks and their sometimes lingering ionization trails, ponder the fact that these meteors are debris from what has been called the single most dangerous object known to humankind. That is Comet Swift-Tuttle, which last was seen at its 1992 return. Experts are now convinced that this comet's large nucleus will not hit Earth at any time in the next few thousand years. But if the nucleus ever does hit, it would probably wreak greater destruction on Earth than the comet or asteroid which destroyed the dinosuaurs and a majority of species on Earth 65 million years ago.

For now—and at least millennia to come—we can enjoy the Perseids without fear of their parent comet. And the Perseids are only part of the delightful experience of August nights. These nights become dramatically longer than those of June and July. Although some or even much of the month may still bring heat and haze, at midnorthern latitudes stronger and more frequent cold fronts typically start passing through. If a few August nights get a little brisk, stargazers will not complain because their skies will then be clearer and starrier. When I watch the Perseids from my home, I sometimes hear from the forest the mournfully beautiful cries of screech owls (what an unfair name!). The choruses of the crickets are loud—they are loud at this moment as I, in mid-August, proofread this chapter one last time!—but becoming more irregular in chant as summer nears its end. August nights are lovely in themselves but to me this is also a wonderful month of anticipation. All the beauties of fall foliage colors and bird migrations and cooler, clear weather will be coming soon to the daytime. And at night, especially late as the Perseids begin to increase in number, I see constellations like Cassiopeia, Perseus, and finally even the bright star Capella come up, heralding the evenings of autumn that are approaching.

Chapter 12

September

[Constellations covered: Delphinus, Sagitta, Vulpecula; Capricornus; Cepheus]

Se, for many of us, the clearest month of the year for stargazing.

The period from late August through much of October offers the greatest number of clear days and nights for most of the United States, and probably also for many other midlatitude countries. September is the middle of this period of least cloudiness. It is also, in my experience in the eastern United States, a time for not just cloudfree nights but some of the very clearest nights of the year. That is to say, it is a month with some very transparent air masses. Certainly cold fronts become stronger and more frequent as we move from August to September and these act to clear out the atmosphere. There can be spells of warm, golden hazy weather in September. But, for the most part, weather changes rapidly and the air doesn't have a chance to stagnate as it often does in summer.

September is a month of changes but they are unlike those of March, for the changes are not thunderstorms, tornadoes, blizzards, and other violent phenomena. Most of September's changes are gentle ones to final resolution and completion—like the completion of the growing season and harvest, watched over by the Harvest Moon's big, bright faithful return to the sky for a number of dusks in a row. Other of September's changes seem more drastic but still express themselves in orderly (in fact, surprisingly and intriguingly orderly) alterations like the migrations of birds and the leaves beginning to change color. These first signs of autumn, in themselves beautiful and eventful, are also moving for several reasons. They are moving because they are not only short-lived but also speak of the farewell and fading of the natural world until spring. As such, they can't help but remind us of the fleeting quality of our own experiences and lives. However, if we ponder more deeply we can also see these events and sensations of autumn as part of a recurring and reassuring pattern of the year—the year which dies but is always renewed. And if on some level we ever doubt the continuity, regularity, and cyclic structure that is so strong in nature, the stars and their seasonal progression are there to remind us.

Actually, a beginner first looking closely at the progression of the seasonal constellations might think the opposite at this time of year: that the progress had been suspended! What I mean is this: If you go out at nightfall in September, you will find that the arrangement of the constellations looks almost the same as it did at nightfall in August. What is happening? The explanation is that nightfall itself is coming earlier in September than in August. In fact, it's around the equinoxes that the time of sunrise and sunset, nightfall and daybreak, change the most rapidly with each passing week. And around the autumn equinox the time of sunset and nightfall is coming rapidly earlier and earlier. So pronounced is this effect that it almost seems to halt the seasonal progression of the constellations. Try this test, however. Go out at the same *clock time* around the same date in August and September (therefore about thirty days apart). You'll see then that the constellations have indeed appeared to move substantially westward. Nothing stops time and the great indicators of it like the Earth moving around the Sun in its orbit.

September is the month when the Northern Hemisphere experiences the autumnal equinox, after which nights become longer than days. Though almost all of us are lovers of sunlight, the part of us that is stargazers welcomes this expansion of night—the domain of the stars. We are happy for nightfall to come earlier and bring us stellar views. And if the earlier nightfalls give us a prolonged chance to enjoy the August wonders of Sagittarius in the south and Cygnus overhead, that is wonderful. There is so much to see in those constellations that one prime month for them is not enough. September is the last month for proper views of southerly Sagittarius before it starts setting too soon after the Sun. Cygnus, on the other hand, takes such a long high journey across our sky that it—and the Summer Triangle—remain well placed for observation in the early evening far into autumn.

Indeed, in this chapter we first visit what might be regarded as the final constellations of summer—the little ones in and near the Summer Triangle we didn't have time to cover in the previous busy months. Delphinus, Sagitta, and Vulpecula: all are small and rather faint, but hold treasures—including one of the most spectacular of all deep-sky objects. And after we tour these little patterns and their wonders, we head north to another fairly dim constellation—this one veritably brimming with amazing deep-sky objects, some of which are unknown to many amateur astronomers. This northern constellation is Cepheus. And when Cepheus is highest, very high in the north, another constellation is at its highest fairly low in the south: Capricornus the Sea Goat. Capricornus is also dim but it offers a few remarkable double stars and its role as a member of the zodiac lends it much interest. It also happens to be an unusually ancient constellation with a strange background in myth, for this constellation is pictured as being half-goat and half-fish.

DELPHINUS, SAGITTA, AND VULPECULA

Two of the three dim little constellations in and around the Summer Triangle of Vega, Altair, and Deneb feature such compact and interestingly geometric patterns that they are quite noticeable if your sky is fairly dark. The pattern of Sagitta the Arrow really is a little arrow composed of stars of magnitudes 3.5 (the arrow's point), 3.7 (shaft and start of feather), 4.4 and 4.4 (the two stars which mark the back of the feather). The pattern of Delphinus the Dolphin is a slightly brighter and definitely more conspicuous little gem: gem indeed, for its main stars form a tiny diamond (of magnitudes 3.6, 3.8, 3.9, and 4.4)—but with a final magnitude 4.0 star forming a tail when you extend a line from the diamond shape that is meant to represent the Dolphin's body. Delphinus seems to be leaping out of the dark water of the night sky just east of the Summer Triangle and the bright band of the summer Milky Way. Sagitta is within the angle of the Summer Triangle which comes to a point in Altair. In fact, it appears to be just breaking

Figure 12.1. The Summer Triangle of Vega, Altair, and Deneb with the summer Milky Way and the Great Rift, photographed by Akira Fujii. The so-called Summer Triangle is actually highest around midevening in September. Can you find the tiny patterns of Delphinus and Sagitta on this image?

through the line between Deneb and Altair and heading east-north-east (the body of Delphinus is pointed northeast).

Delphinus the Dolphin

Let's consider Delphinus first. Its pattern is conspicuous enough to have earned it a name in lore which no one has been able to under-stand or trace back to an original source: Job's Coffin (there is cer-tainly nothing in the biblical story of Job about his coffin). The Greek myth associated with Delphinus is that of the great musician Arion. Arion was robbed and forced to leap overboard by a crew of thieving sailors who intended to continue on to his home city and tell of the tragic accident by which he died. Instead, dolphins that had heard Arion's songs from the ship rescued Arion and one of them bore him to the city faster than the ship. When the ship arrived and the crew had told their tale, Arion suddenly stepped out. We can only imagine the looks of shock on the faces of the murderous crew when they saw Arion alive and were hauled off by the authorities to face their just punishment.

Delphinus offers one beautiful and easy double star—or, rather, a close-together pair of doubles. The main attraction is Gamma Delphini, the tip of the Dolphin's nose. It consists of a 4.5 and 5.5 star that are 10″ apart. There is some dispute over the colors and some people see them both as yellowish white or even white. But I know that I, and perhaps most people, perceive the brighter star as yellow-white and the dimmer star as having a delicate hint of green. It's important not to use too much aperture on such a bright double or you will wash out the colors. A 4-inch telescope will do fine and an 8-inch may be almost the largest you would want to use. Gamma Delphini is one of the finest double stars of its kind in all the heavens. But James Mullaney points out that it is even more interesting because it has a "Ghost Double." This Ghost Double is only 15′ to the southwest of Gamma Delphini and is formed by a 7.6 and 8.4 star that are 6″ apart. Mullaney notes that both of these doubles lie at about 100 light-years from Earth but are heading through space in nearly opposite directions![13]

The origin of the proper names for Alpha and Beta Delphini is amusing. The stars started appearing on lists and star atlases in the nineteenth century as Sualocin and Rotanev. But scholars could neither trace the origin of those names nor their meaning. It turned out that the astronomer Giuseppe Piazzi had mischievously reversed the Latin version of his assistant's first and last name and applied them to the two Delphinus stars. The assistant's name in Italian was Niccolo Cacciatore. In English that would be Nick Hunter. In Latin it is Nicolaus Venator. And those two words spelled backward are Sualocin and Rotanev.

Sagitta the Arrow

Sagitta the Arrow could be any famous arrow of myth or legend. That's just the problem: Which one is it? If there was a nearby constellation which seems to be shooting it, then Sagitta could be attached to that constellation's myths. Well, of course, there is Sagittarius the Archer—which with another constellational centaur, Centaurus, is sometimes said to be the wise Chiron, teacher of several of the greatest heroes of Greek myth. But Sagittarius is rather far from Sagitta in the sky and when depicted as a centaur has always been shown facing (and preparing to shoot) west, possibly toward Antares, the heart of Scorpius. Sagitta is flying in almost the opposite direction. There is, however, a constellation neighboring Sagitta which might be the shooter of this arrow: Hercules, constellation of the legendary strongman. That constellation, as we saw in the June chapter, is upside-down in the present sky of midnorthern latitudes, and is portrayed as kneeling on one knee. But he is sometimes portayed as kneeling on one knee in the act of firing (or having just fired) an arrow from a bow. His target? Possibly Cygnus, Lyra, and Aquila. All three have been considered birds at one time or other: Cygnus the Swan and Aquila the Eagle still are, and Lyra the Lyre has been pictured as a vulture in the past. But in this imagining they would be three of the same kind—the evil Stymphalian birds which Hercules was trying to drive away. If Sagitta is one of his arrows, it appears to be one that has missed hitting any of the birds.

The most notable deep-sky object in Sagitta is M71. This globular cluster is conveniently located on the shaft of Sagitta the Arrow, about midway between Gamma Sagittae and Beta Sagittae. When you hear that M71 shines at about 8.4 with a diameter of about 7′ you might think it is not one of the most impressive globular clusters to observe. Actually, medium magnification on a medium-size telescope begins to resolve this cluster impressively into numerous stars but—and this is the important point—without any real central condensation. For a long time, this led to opinions that M71 was actually a very rich open cluster. Astronomers have proven, however, that M71 really is a globular cluster—just an unusually close (13,000 light-years distant) and loosely organized one. This object is interesting in small telescopes but really only becomes truly impressive when seen in a fairly large amateur instrument.

Vulpecula the Fox

Vulpecula the Fox is dimmer than Delphinus and Sagitta and has no prominent pattern, merely filling with dim stars the strip of sky between them (the Dolphin and Arrow) and Cygnus. But this last of constellations in alphabetical listings has at least three easy and exciting deep-sky objects—one of them among the heavens' greatest showpieces.

One of the three sights is an open cluster and the other is an asterism. The former is NGC 6940, a rich half-degree-wide, magnitude 6.2 open cluster. The other is an asterism which could hardly be called rich but is almost certainly the most striking of all star patterns in a small telescope at very low power. I refer to the grouping variously called Collinder 399, Brocchi's Cluster, and, most commonly, the Coathanger.

The Coathanger measures about a degree long and at a total magnitude of 3.6 it can be glimpsed as a fuzzy patch with the naked eye in a good dark sky. Even weak binoculars reveal its secret, though the view is perhaps best in a rich-field telescope at very low power. What you see is an almost perfect line of six similarly bright stars with a hook of four more stars sticking out from the center of the line. It's astonishing. I still remember the first night that I saw this object. I was a teen scanning around the Summer Triangle region with a small telescope. The Coathanger was not very well known in those days but I had heard of it. It's a good thing that the memory of the name had stuck with me in the back of my mind because when it suddenly entered my telescope's field of view I thought for a second that I had lost my wits or was the victim of some eerie cosmic joke. Fortunately, it only took me a few moments to remember having heard of an asterism called the Coathanger and realizing that this must be it.

Actually, until a few years ago there was still doubt whether the Coathanger was a true cluster—Brocchi's Cluster—or was mostly just an asterism, mostly a coincidental lineup of physically unrelated stars. The latter is true.

You can find the Coathanger roughly a third of the way along the line from Altair to Vega.

Our final object in Vulpecula is not quite as famous or frequently observed or fondly thought of as M57, the Ring Nebula in Lyra. But M27, the Dumbbell Nebula, is actually much bigger and brighter than even the Ring Nebula; in fact, M27 is easily the most prominent of all planetary nebulas.

M27 is big. It measures 8' by 5', so its length is about one-quarter the apparent diameter of the Moon. The biggest planetary, Helix Nebula, is bigger—much bigger—but its light is spread out much too widely and therefore appears as an elusive low-surface-brightness ghost. M27 shines at magnitude 7.6 and at its size this is bright enough to make it quite intense, capable of bearing rather high magnification without the apparent surface brightness getting too low.

The Dumbbell is so named because of its shape, which in small to medium-size telescopes does at first glance resemble a dumbbell with two roundish lobes and a narrower zone connecting them. But M27 actually looks more like an hourglass or a "celestial pillow floating serenely among the stars"[14]—or like an apple core. The fascinating thing is that in really clear, dark skies with medium to large amateur telescopes the eaten apple begins to fill back in with faint luminosity! The central star of this nebula released its shells of gas a few tens of thousands of years ago. It is estimated to shine at about magnitude 13.9 but is much harder to see than that suggests because of the surrounding nebula's brightness. At the tip of one of the projections which stick out from both sides of both ends of the "apple core" is a much brighter star that is quite easy to see even in a medium-size telescope. And the entire surroundings of M27 are beautifully rich with stars.

M27 warrants long periods of observation, especially at higher powers with fairly large telescopes. Can you see color in it? As with the Ring Nebula (discussed in the July chapter), I have glimpsed several colors in the Dumbbell.

There's no doubt that M27 is worth many an admiring observation. But how do you locate the big planetary? I've often scanned by telescope from Delta to Gamma Sagittae—that is, along the shaft of the Arrow—and then made a right-angle turn (more or less northerly) and traveled about the same distance to find the Dumbbell. If you have an equatorial mount, the task is simpler because M27 lies 3.3° almost due north of Gamma Sagittae. When you get there, you'll find that M27 is less than 1/2° south of 5.7-magnitude 14 Vulpeculae.

CAPRICORNUS

A considerable distance due south of Delphinus and eastern Vulpecula is a rather dim constellation whose stars form a pattern that is hard to describe. It also contains unusually few deep-sky objects, for it is quite a considerable distance east of the Milky Way band—in fact, of that part of the band which blazes so brightly and with such rich adornment of nebulas and star clusters near the Teapot of Sagittarius. However, its great claim to fame is that it is a member of the zodiac, the next constellation the Sun, Moon, and planets enter after they leave Sagittarius. It is Capricornus the Sea Goat, or Goat-Fish.

The creature portayed by these stars is depicted as having the front half of a goat and back half of a fish. Where could such an idea have arisen? Centaurs may have first entered human consciousness when a culture without horses saw another with men riding on horseback. One explanation for how unicorns came into being is that travelers to Africa saw rhinos and brought back accounts of a creature with a horn on its nose. But why would a goat and a fish get conjoined into one creature in myth? We don't know, but the fact that there was one god—the Sumerian god Ea—who was pictured in this form seems strong evidence that Capricornus was associated with Ea and that these stars first took this form when Ea was popular—maybe more than four thousand years ago. The strange creature must have become so strongly associated with these stars that later cultures like the Greeks and Romans simply adopted it for their Capricornus, though "Capricornus" in Latin simply means "horned goat."

Even today we must admit that it is hard to find any one common animal or thing which much resembles the main pattern of the stars of Capricornus. Guy Ottewell says it looks more like a boat than a goat. But if so, it is like a little paper boat. I have said it reminds me of a (rather stylistic) origami bird! Or perhaps it is the kerchief of a Western outlaw pulled up over the lower part of his face?

Fortunately, to find the two most interesting deep-sky objects of Capricornus you don't really need to have in mind a clear image of the entire pattern of the constellation. For at its western, leading end is a sight to which your naked eyes or binoculars may be drawn. To the unaided vision, there are two moderately bright stars only a few degrees from each other, with the upper star itself a close naked-eye double. The close naked-eye double is Alpha^1 and Alpha^2 Capricorni. These are, respectively, a magnitude 4.3 and 3.6 star that are about 6 1/2' apart. That is much closer than Mizar and Alcor (see the chapter for April), but the two stars are a lot more similar in brightness, a factor which aids in the eye seeing both of them. Do you see these stars as yellow-white or with a touch of orange? Each has a faint companion (9th- and 11th-magnitude, 7" and 45" separated). Alpha^2 and Alpha^1, which are known collectively as Algiedi, are merely an optical double. In other words, they are not physically related to each other but simply appear along nearly the same line of sight to us. Alpha^2, the star of greater apparent brightness, is 108 light-years from Earth, while Alpha^1 is about 700 light-years from us.

Now turn your binoculars on the Alpha^1/Alpha^2 pair and the star a few degrees south of it. That star is Beta Capricorni (Dabih) and magnification reveals that it, too, is a wide double star—the components just under 3 1/2′ apart. In this case, though, the components are magnitude 3.1 and 6.1, so the dimmer star is too faint (and too much fainter than the primary star) for Beta Capricorni to be seen as a pair by even the keenest naked eye. In binoculars, of course, it is an easy and attractive sight. And telescopes show the yellow and blue or orange and blue hues.

Alpha and Beta Capricorni mark the head of the Goat-Fish. Interestingly, two stars similarly near each other mark the opposite end, the eastern end of the constellation—the tail of the fish part. These are Delta Capricorni (Deneb Algiedi) and Gamma Capricorni (Nashira), magnitude 2.9 and 3.7, respectively. Algiedi is derived from an Arabic word meaning "the goat," and Deneb Algiedi is "the tail of the goat." Delta Capricorni is located only 38.5 light-years from Earth. By the way, the planet Neptune was discovered about 4° northeast of Delta Capricorni, near Mu Capricorni. The discovery was made on September 25, 1846 by J. Galle based upon the calculations of U. J. Leverrier. Also calculating Neptune's position before discovery was J. C. Adams. The interesting thing is that as the twenty-first century begins, Neptune is again in Capricornus. The planet will pass its discovery position, completing its first full observed circuit of the sky (its orbit is 165 years long) on July 21, 2011.

Do you need help in finding the main part of Capricornus between its two interesting two-star ends? Or even in locating Alpha and Beta Capricorni? Here's a neat trick I've found: if you take the line from Vega to Altair and extend it one more length of itself, it will take you to the pattern of Capricornus.

By the way, there is one Messier object which does lie within the boundaries of Capricornus. The globular cluster M30 shines about magnitude 7 1/2 and requires a fairly large amateur telescope to start resolving. It is often overlooked because it is rather southerly and a full 6 1/2° east-southeast of magnitude 3.7 Zeta Capricorni, the lower left star of the inconspicuous Capricornus pattern. M30 does lie only 1/2° west-northwest of 5.5-magnitude 41 Capricorni.

CEPHEUS

Now we journey to another rather dim constellation, one which on September evenings is very high in the northern sky for observers at midnorthern latitudes. Cepheus the King can be found in a dark sky by continuing northeast up the Milky Way band from Cygnus. We find that the Milky Way here is much less brilliant than in the Cygnus Star Cloud but actually still fairly bright and very complex and interesting. The main pattern of Cepheus is offset from the center of the Milky Way band and even a veteran constellation watcher may need a moment to get oriented enough to trace the shape, which is like that of a house with a steep roof. Often a first step in finding it is locating the King's much brighter mate, Cassiopeia. There is a little triangle of stars about midway between Deneb and the bright zigzag of Cassiopeia and this marks the southeast corner of the "house," part of its bottom or foundation. On September evenings facing north, the house appears upside-down. The point or peak of the roof is Gamma Cephei (Errai), which is only about 13° from the current north celestial pole and Polaris. This magnitude 3.2 star, only 45 light-years away, will take its turn as the North Star about two thousand to twenty-five hundred years in the future.

The brightest star in Cepheus, which forms the southwestern corner of the house, is magnitude 2.5 Alpha Cephei (Alderamin), which will be a fine North Star around the year 7500.

Now let's return to the southeastern corner of the Cepheus house, to that little triangle midway between Deneb and Cassiopeia, in the midst of the Milky Way band. The most famous star in Cepheus in modern times is a member of that little triangle, the member nearest to Cassiopeia. This is Delta Cephei, the prototypical object of the very important class of variable stars named for it, the Cepheids. Delta Cephei shines as bright as magnitude 3.5, the tiniest bit dimmer than the other bright member of the triangle, Zeta Cephei. But during the course of about four days, it fades to 4.4, a bit dimmer than the dim member of the little triangle, Epsilon Cephei. Then, in only about one and a half days, Delta brightens back up to its maximum. The whole period is precisely five days, eight hours, and forty-eight minutes. There are brighter Cepheids than Delta (Polaris itself is one, with an extremely small brightness variation) but no brighter one shows a dramatic range. The star's variability was discovered in 1784 by the deaf-mute astronomer John Goodricke.

The most important characteristic of the Cepheids is that they display a *period-luminosity relation*—that is, a proportion between the length of their period and their luminosity. The longer the period, the more luminous the star. This has proven an absolutely invaluable means for measuring distances, for if the period of a Cepheid tells us what its luminosity (its true brightness) is, and we observe its apparent brightness, then we can figure out how far away it must be to look that bright. Delta Cephei is about 1,000 light-years from Earth. As more powerful telescopes have been able to find Cepheids in galaxies other than our own, they have enabled us to determine the distances to those galaxies. The period-luminosity relation, while not quite as simple to use as suggested here (there are several different types of Cepheids, for instance), has been so valuable that it truly warrants the Cepheids being called "yardsticks of the universe."

The performances of Delta Cephei can be easily followed with the naked eye and the fame of it as a variable star has overshadowed the fact that it is also, in the telescope, a gorgeous double star. The yellow or orange variable has a blue companion that shines at magnitude 6.3 an easy 41″ away. There are several other fine doubles in Cepheus, including Xi Cephei (magnitudes 4.4, 6.5, 8″ apart), in the middle of the "house" of Cepheus.

But we need to return to the little triangle in which Delta Cephei shines. Draw the line from the triangle to brighter Alpha Cephei, and find the star little more than a degree southwest of the middle of this line. A good pair of binoculars or small telescope will help confirm the star, for it is one of the reddest of naked-eye objects: Mu Cephei, also known as Herschel's Garnet Star.

This is the brightest and probably the most famous of all the heavens' "very red" stars. Its brightness ranges typically from about 3.9 to 4.5, and, although the changes are irregular they are said to follow an average period of 755 days, with smaller variations noticeable over periods of one hundred days or less. Extreme brightness values have been 3.4 and 5.1. Mu Cephei looks almost truly red at times in a three-inch or four-inch telescope but only deep orange in an eight-inch. This huge star is so cool that ordinary steam has been detected in it by its spectrum. It is located about 3,000 light-years away from Earth.

About 2.5° south of Mu Cephei is a lovely triple star inside a big sparse open cluster and huge but very faint and elusive nebula. The triple star is Struve 2816 which consists of a very hot O6 star of magnitude 5.6 with magnitude 7.7 and 7.8 stars 12″ and 20″ from it. Use

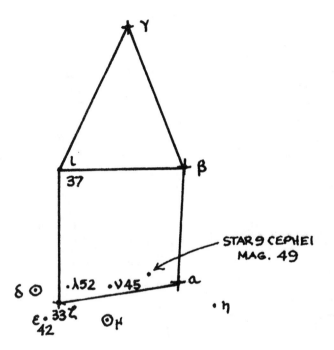

MU (μ) CEPHEI

Figure 12.2. Diagram showing reddish variable star Mu Cephei and comparison stars of unvarying brightness. Numbers are magnitudes of the stars with the decimal points removed (for example, "37" means magnitude 3.7).

binoculars to enjoy the 50'-wide grouping of stars combining to a total magnitude of 3.5 around this triple. In very dark skies or with a nebula filter, binoculars as small as 7 × 50 can show traces of the huge (170' by 140'!) nebula IC 1396 centered around the cluster. The north edge of the nebula stretches almost all the way to Mu Cephei.

HARVEST MOON AND THE MOON ILLUSION

We turn from our September stellar travels to appreciate another phenomenon of the skies at this time of year. The Full Moon nearest to the autumn equinox usually falls in September and it bears a famous title: Harvest Moon. Songs have been written about it and the name evokes a rich tapestry of early autumn sights and experiences. But Harvest Moon is more than just a pretty, potent name.

Around the autumn equinox the parts of the ecliptic and zodiac which is rising at nightfall makes its shallowest possible angle with the east horizon. What this does is make the Moon rise at "almost the same time" for a number of nights around Full Moon as seen from midnothern latitudes. Actually, around 40° North "almost the same time" is not much less than thirty minutes later each night for a small number of nights. But the *average* delay of moonrise during the year on succeeding nights at this latitude is more than fifty minutes. And the farther north of 40° N you go, the shorter the interval between successive moonrises at Harvest Moon gets.

This behavior can be a bit troublesome for stargazers who want to see the bright Moon start rising later in the evening so that they can get in some observing of faint stars and deep-sky objects in the hours after nightfall. But seeing the big, bright Moon coming up like an old and reliable friend around the same time in dusk for night after night is a marvelous experience. Also, the Harvest Moon phenomenon was more than just beautiful and interesting to farmers in days gone by. It was very useful, sometimes perhaps even critical. For the rising of a bright Moon just as daylight started fading allowed farmers to continue harvesting their crops for longer each day, if they needed to.

A number of cultures have celebrated the Harvest Moon. One of them is the Chinese, for whom this Full Moon is marked by the important and joyous holiday called the Midautumn Festival or Festival of the August Moon. "Midautumn" here is from an older concept of autumn which places its middle at the autumn equinox. And the reference to the August rather than the September moon is due to a discrepancy in relating the Chinese calendar to the Western calendar.

One reason that the Harvest Moon is marvelous to look at is that any rising or setting Moon—any Moon low in the sky—can seem to look far bigger than it does when it is higher in the sky. Yet we know that the Moon is actually *closer* to the observer when high in the sky: Earth has then rotated the observer around to bring the observer nearer to the Moon by about one Earth radius. So what is the cause of the so-called Moon Illusion?

Scientists are still trying to figure out the complete set of elements which enter into producing the illusion. But it is not just begging the question to say that the illusion depends on the observer's unconscious and indirect perception of the shape of the sky. The perceived shape of the sky is different under different conditions. For instance, it depends on how uniformly the sky is lit (more uniformly on an overcast day than clear, etc.). When does the sky seem like a perfectly rounded hemispherical dome to us and when does it seem much more flattened, with the zenith near above us and the sky down by the horizons very far? Certainly one thing which makes the low Moon (or Sun or a very bright constellation like Orion) appear larger when it is rising or setting is that we unconsciously are comparing it to distant and therefore small-looking objects in the landscape. If we have a view of moonrise unobstructed by any close trees or buildings, then we will see the Moon come up next to a distant tree or building, which is small in angular size and therefore by comparison makes the Moon look much bigger. When the Moon is high in the sky, we seldom have any familiar landmarks near it to compare it to—and if we do, they tower near above us and look big, making the Moon look small.

This comparing of the Moon to distant and therefore small-looking objects in the landscape is by no means all there is to our perception of the shape of the sky and the production of the Moon Illusion. For one thing, there are certain nights when the Moon looks bigger than others—on these occasions, some observers estimate the low Moon as appearing two to three times wider than when it is high.

Many people tend to associate the Moon Illusion with late summer and early fall. There are several reasons why this time of year—and especially September—is when the most sightings of the Moon Illusion are made. First of all, this is the most cloudfree time of year for most of us. Perhaps even more importantly, the Harvest Moon phenomenon now brings up the nearly Full Moon at dusk on a number of nights, and September dusks occur at just about the right clock time and temperature for many people to be outside to view them.

SEPTEMBER'S OBSERVING EXPERIENCE

We've already touched upon many of the beautiful ingredients of the observer's surrounding natural and sensory environment on September evenings. The big Harvest Moon one week of the month.

Another week or two of the month, some of the starriest skies you'll ever see, courtesy of air frequently moved and freshened by cold fronts. Several of the very clearest nights I've ever experienced occurred in September, including one night I saw a star as faint as magnitude 7.9 from near sea level with my naked eyes.

Of course, there are usually exceptions to rules. For people living around the Gulf Coast and up the Eastern seaboard, September is the most active month for hurricanes and tropical storms. For most of the United States, this is the best time of year for a local astronomy club to plan a weekend-long star party since you have the best chance of getting one or more of your nights clear. But once in every so many years a tropical cyclone or its remains will directly or indirectly give you a rainout.

For the most part, though, September weather is excellent for astronomy. The only other problem is that diurnal (daily) temperature ranges may be great with strong radiational cooling, leading to fog or to dew forming readily on optics. This will not happen on the driest, clearest nights or if the wind is blowing in the hours right after a cold front passage. You can also try observing partly under the shelter of trees. But if you are going to observe out in an open field where you can see most of the heavens, it is worthwhile to invest in a "dewcap" (extended cylinder to attach to the sky end of your telescope to keep the dew off the optics), or even an apparatus for gently heating your optics (such heating systems are commercially available).

From the practical, I turn back to the aesthetic and spiritual aspects of observing on September nights. To me, perhaps the most moving experience of nights in late September and early October is to be out in the crisp air under a sky almost inexpressibly rich and bright with stars and then to hear a sound: the haunting call of the migrating geese. As a flock passes slowly, and usually invisibly, overhead or across one of the sides of the sky (it is often mystically difficult to tell where), the individual cries of the birds become distinguishable. Most species of North American birds migrate at night and many rely at least in part upon specific star patterns to help them navigate (this has been proven in experiments with birds in planetariums). Few sounds are more evocative than those of the geese, usually heard high, sometimes thrillingly low, heeding their age-old call to head south to their warmer wintering grounds. Sometimes it seems to me that we human beings are drawn by a similar instinct: a powerful urge to seek and watch the stars.

Chapter 13

October

[Constellations covered: Pegasus; Equuleus, Lacerta; Aquarius, Piscis Austrinus; Pisces; Cetus, Sculptor]

"**O**ctober is the month of painted leaves," wrote author and naturalist Henry David Thoreau. Few places in the world rival his native New England for the beauty of its autumn foliage. Throughout many northern lands, however, October is a time of orange or golden leaves against blue skies by day . . . and clear, cool weather to enjoy crystal-clear visions of the heavens during the lengthening nights.

Strangely, though, this month when the deciduous forests of the Northern Hemisphere flame at their brightest is the month when a vast stretch of the night sky is filled with dimmest constellations. In the middle of October evenings, the lower half of the sky all the way from the southwest to the east is so sparsely starred that the sight is actually dramatic in its starkness. This entire vast, lonely stretch of sky is broken by only two 2nd-magnitude stars and one low 1st-magnitude beacon—the star Fomalhaut in Piscis Austrinus.

The southwest-to-east span of heavens has so few prominent naked-eye stars that it seems awash in an ocean of darkness. And so it is appropriate that this entire "ocean of darkness" is occupied by constellations related to water and has been called the Water or the Great Celestial Sea. At least five constellations, including a few of the largest in all the heavens, are found here. The first one of these aquatic constellations in the night's progression, Capricornus the Sea Goat, has already passed the sky's central meridian by the middle of October evenings. We examined Capricornus in our previous chapter. But following Capricornus—and at their highest in the south this month of October—are Aquarius the Water Bearer and Piscis Austrinus the Southern Fish. Following Aquarius and Piscis Austrinus across the southeast sky are Pisces the Fish and Cetus the Whale.

We could add as marginal members of the Water a constellation that we studied last month, Delphinus the Dolphin; the very southerly constellation below Piscis Austrinus, Grus the Crane (walking on stilt legs at the shore of the Water); and Eridanus the River, rising in the east and southeast under Cetus. But our southern focus in this chapter will be Aquarius, Piscis Austrinus, Pisces, and Cetus—the central, certain members of the Great Celestial Sea along with Capricornus.

Notice I say "southern focus." For there is another focus in this chapter, a focus on one other constellation we will examine, one higher in the sky, above the Water. This rather bright constellation is famous for its usefulness in locating others. It is also the first of a great group of bright constelaltions which are like a brigade of brightness already marching up the northeast sky: Andromeda, Cassiopeia, Perseus (even the early winter constellations Taurus and Auriga are already visible low in the east and northeast sky by the middle of October evenings). We will be saving them for next month when they are higher. But this month we will take a close look at the first member of this parade of brighter patterns: Pegasus the Winged Horse.

PEGASUS

Pegasus is perhaps the most famous creature in Greek mythology. The flying horse leaped up from the blood of Medusa and so is part of the great myth of Perseus, which I related in the December chapter. But Pegasus was later the steed of the hero Bellerophon, who among other feats killed the monster called the Chimera. Unfortunately, Bellerophon eventually became vain and decided that he would ride Pegasus right up to Mt. Olympus where the gods dwell. But the gods sent a fly to bite Pegasus on the journey up. The horse bucked and threw Bellerophon, who fell to his death. Pegasus himself was fault-less and so earned his own constellation.

This marvelous horse is, fittingly, a marvelous and important con-stellation. Pegasus soars very high in the south for viewers at mid-northern latitudes on October evenings. Strangely, it is depicted flying upside-down. The nose of the horse is the magnitude 2.4 orange star Epsilon Pegasi, also known as Enif (from the Arabic for "the nose"). Enif shines approximately two hours of right ascension east—almost exactly east—of Altair. The horse's neck is formed by a southeast-to-northwest oriented line of stars that ends with magnitude 2.5 Alpha Pegasi (Markab). Alpha Pegasi forms the body of the horse with three other stars of similar brightness. The pattern of the four stars is so symmetrical and interesting that it has earned its own name as an asterism: the Great Square of Pegasus.

The Great Square

The Great Square of Pegasus is somewhat longer in the east-west direction than it is in the north-south, so it is really a rectangle. That does not alter the fact, however, that this pattern is highly useful for finding other constellations and stars. All lines of sight drawn from the sides of the Great Square are interesting.

For instance, a line drawn down (south) from the the west side of the Square, if extended most of the way toward the horizon directs the eye to 1st-magnitude star Fomalhaut. A line drawn down the east side of the Great Square and extended passes near 2nd-magnitude Beta Ceti. What's additionally interesting about the west and east sides of the Square is that they are almost exactly north-south lines *and* follow closely the 23h and 0h lines of right ascension. Actually the two stars which form the east side are at 0h 13m and 0h 8m—a little east of the 0h line of right ascension (so the Great Square is a bit more than one hour of right ascension wide, west to east). But the stars of the east side still serve as excellent guides to important stars farther north which in turn, help point the way to the north pole: a line extended north from the east side of the Great Square takes us to Beta Cas-siopeiae, the lead star of the "letter M" pattern of Cassiopeia, and Gamma Cephei (which we met last month as the peak of the roof of the little house pattern formed by the brightest stars of Cepheus).

Let's note the names and magnitudes of the stars of the Great

Square. If we go clockwise and start with the northwest star, the roll reads as follows: Beta Pegasi (Scheat), magnitude 2.5 (but slightly variable, with ultimate extremes of 2.3 and 2.7); Alpha Pegasi (Markab), magnitude 2.5; Gamma Pegasi (Algenib), magnitude 2.8; and Alpha Andromedae (Alpheratz or Sirrah), magnitude 2.1. The last star—the one which marks the northeast corner of the Great Square— is technically a member of the neighboring constellation Andromeda. (This custom of two constellations sharing the star of one, so that both can form their patterns, is also practiced with the early winter constellations Auriga and Taurus.) By the way, what could have been the stars of the back legs of Pegasus are the stars of the head and body of Andromeda the Chained Maiden. Pegasus has to be pictured without his back legs—perhaps he is leaping up from the sea and has not entirely emerged.

The Great Square can be used as a rough test of how dark and clear your sky is at a given site and night. From a suburban or small city location, you may with the naked eye see only two faint stars within the rather large area of sky bounded by the Square. From a really dark country location where you have a naked-eye limiting magnitude of 6.5, you can detect about three dozen faint stars within the Square!

Pegasus Deep-sky Objects, Plus Equuleus and Lacerta

Pegasus has relatively few easy deep-sky objects. There are no dramatic double stars or open clusters here. There are, though, some challenging galaxies. Least difficult is magnitude 9.5 spiral NGC 7331, which measures about 11′ by 4′. Only 30′ to its south-southwest is a famous group of galaxies called Stephan's Quintet. If you've ever seen the movie *It's a Wonderful Life*, you've seen a professional astrophotograph of this group. These galaxies are well beyond the range of small telescopes, beginning observers, and mediocre skies. At least a 10-inch telescope and superb skies are needed to glimpse these 14th- and 15th-magnitude objects. But if you can locate and see NGC 7331, you will find it interesting to think that the famous Stephan's Quintet, though too faint to see, is in the same low-power field of view.

Pegasus may have few easy deep-sky objects but one it does have is among the finest globular clusters in the heavens—M15. M15 lies about 4° northwest of bright Epsilon Pegasi, right in line with that of the head of Pegasus formed by Theta Pegasi and Epsilon. Perhaps Pegasus is sniffing for M15—as you will be when you learn that it shines at about magnitude 6.3 and is as much as 18' across. A 6th-magnitude star sparkles close by. Along with its near-twin M2 (which we'll discuss in a moment), M15 is probably tied for fifth in brightness and overall spectacle in the rankings of the globular clusters properly visible from midnorthern latitudes. Only M13, M22, M3, and M5 are brighter and more glorious. But each of the great globular clusters is unique and has its own merits.

EQUULEUS AND LACERTA

After you have observed M15, try looking for some dim naked-eye stars just west of the head of Pegasus. Those stars form a pattern no bigger than the head of Pegasus—the pattern of Equuleus the Little Horse or Foal. Equuleus is one of the dimmest constellations. Its brightest star is Alpha Equulei, which shines at magnitude 3.9 and is also known as Kitalpha. You should also look for another dim little autumn constellation, this one well northwest of the Great Square and just south of the halfway point between the bright patterns of Cygnus and Cassiopeia. It is Lacerta the Lizard, and does contain some attractive deep-sky objects. Among these are the magnitude 6 1/2° open cluster NGC 7209 and the double star 8 Lacertae (magnitudes 5.7 and 6.5, separation 22.4").

AQUARIUS AND FOMALHAUT

As we have already noted, bright Fomalhaut shines far to the due south of the west side of the Great Square. The constellation which fills most of the space between Fomalhaut and Pegasus is Aquarius the Water Bearer. Like Capricornus to its west and Pisces to its east, Aquarius is a dim constellation of the zodiac. Aquarius does, however, have a few deep-sky objects more interesting than those of its neighbors.

Aquarius

First we should trace out what pattern Aquarius does have. The constellation is portayed as a young man pouring a stream of water from a container. This youth is usually identified with Ganymede, the servant of Zeus/Jupiter who also has a moon named after him (the largest moon in the solar system, orbiting the planet Jupiter). The focal point in the constellation is a little asterism, the Water Jar or Urn. The Water Jar looks like a tiny Y tipped over to the left (east). Three of its stars are about magnitude 4.0; the fourth is fainter. The Water Jar is right smack on the celestial equator, as is one of Aquarius's two brightest stars, magnitude 3.0 Alpha Aquarii (Sadalmelik) just a few degrees west of the Water Jar. And quite a few more degrees west on the celestial equator is the great Aquarius globular cluster M2.

M2 is situated less than a degree south of the celestial equator. Remarkably, it lies almost precisely—and about 13°—due south of the very similar other great globular cluster of fall, M15. M2 may be a bit dimmer than M15. Both are unusual in being rather oblate (larger across their equator than pole to pole). With an 8-inch or larger telescope, do you notice color in M2—perhaps more readily than in M15? Compare these two wondrous citadels of stars with each other extensively. You won't have any other globular cluster nearly so good as these high in your evening sky until spring brings M3 and then M13 into your view.

From the Water Jar and Alpha Aquarii, located in the center of the north edge of Aquarius, spray out a series of lines to an assortment of other stars in Aquarius. Some of these lines are rivulets within the stream of water that Aquarius is pouring. Others may be his limbs—for instance, the curving line which extends from Alpha Aquarii to Beta Aquarii (magnitude 2.9 Sadalsuud) and on southwest to magnitude 3.8 Epsilon Aquarii. About 4° southeast of Epsilon (or 10° due east of Alpha Capricorni) is the dimmest globular cluster in Messier's catalog, M72, and a little farther from the star is a weak gathering of a few stars which Messier thought had some nebulosity about it (it doesn't) and dubbed M73. But just a few degrees from M72 and M73, about 1° west of Nu Aquarii, is a far more intriguing object which Messier missed: NGC 7009, the Saturn Nebula.

It's not surprising that Messier missed NGC 7009, for at very low power it looks like an only very slightly fuzzy star. A little more magnification reveals an intense (high-surface-brightness), blue (to some, green) flattened ovoid nebula. An 8-inch telescope is certainly sufficient in excellent conditions to show far more: an outer shell and possibly the projections to either side of the inner nebula which roughly resemble Saturn's rings when they are nearly edge-on. The "globe" and "rings" of the Saturn Nebula are of almost the same size as Saturn's during a good presentation of the planet. At magnitude 8.3, NGC 7009 is one of the very brightest planetary nebulas.

The brightest planetary nebula of all is also located in Aquarius. But NGC 7293, the Helix Nebula, is not only less prominent than the Saturn Nebula, it is positively elusive. The reason is its immense size. The Helix is the closest of planetary nebulas—only about 500 light-years away, six times closer than the Saturn Nebula. And therefore it spreads over 12' by 16' and its overall magnitude of 6.5 is only enough to make it appear as a faint ghostly circular shape if you have a dark sky and use very low magnification. I've gotten good views of the Helix Nebula with a 4 1/4-inch rich-field telescope (without a nebula filter) and it can be seen in large binoculars. A nebula filter is very helpful in making the Helix more visible.

Getting to the location of the Helix is not easy to do on your own. It lies in a deserted section of Aquarius, not even near to the major "strands of water" pouring down the sky from the Water Jar. If you are not using a go-to telescope or setting circles, a possible strategy for getting there is to look a few degrees northeast of the midpoint of the line from Fomalhaut to Delta Capricorni (Deneb Algiedi).

Fomalhaut and Piscis Austrinus

We come at last to Fomalhaut, which at magnitude 1.17 is ever so slightly dimmer than Pollux and therefore the eighteenth brightest star in the night sky. But Fomalhaut is the brightest star of the traditional constellations of autumn. It is notable for being the most southerly of the 1st-magnitude stars visible from 40° N. Though only a few degrees farther south than summer's Antares, Fomalhaut has no bright fellow stars in its section of sky the way Antares does in Scorpius. The constellation in which Fomalhaut resides, Piscis Austrinus the Southern Fish, has no other stars brighter than fourth magnitude and no striking deep-sky object for beginners. Indeed, there are no brighter-than-average stars anywhere near Fomalhaut. The star stands

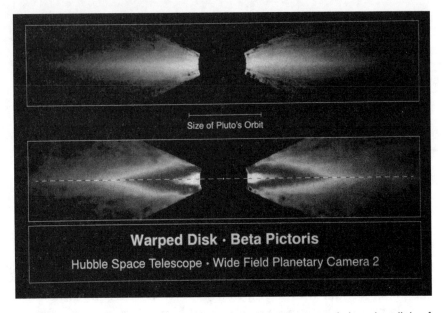

Warped Disk · Beta Pictoris

Hubble Space Telescope · Wide Field Planetary Camera 2

Figure 13.1. This Hubble Space Telescope image proved that the disk of dust around Beta Pictoris is warped, implying the presence of planets in the Beta Pictoris system. In 2002 a similar finding was made about the disk of dust around 1st-magnitude star Fomalhaut.

like a lonely lighthouse beacon near the shore of the Great Celestial Sea of water constellations.

The Hipparcos satellite's measurements enabled us to determine that Fomalhaut lies at 25.1 light-years from us, just 0.2 light-year closer than Vega—but since the margin of error is 0.1 or 0.2 light-year either way, they might be at exactly the same distance from us. Fomalhaut is somewhat less luminous than Vega. It is also a little cooler and shows no bluish tint in its light, displaying only white when it is high enough in a really clear sky not to be reddened by absorption from haze in our atmosphere. Two decades ago, Fomalhaut and Vega were the two bright stars which the IRAS satellite showed to have disks of dust around them—possible belts of comets. Then, in 2002, discovery of a warp in the disk of Fomalhaut indicated that it must have a planet—the 1st instance ever of evidence for the existence of planets around one of our 1st-magnitude stars.

Some people suppose that the name Fomalhaut should be pronounced as if it were French: FOHM-uh-loh. Actually, it should be pronounced FOHM-uh-lawt, being from an Arabic phrase that means "mouth of the fish."

PISCES

The fish of Piscis Austrinus is single—one fish. A far more famous constellation is the next in the zodiac working east after Aquarius and it is called Pisces the Fish—the plural, two fish. Pisces is the current residing place of the vernal equinox point in the heavens—the point on the celestial equator where the Sun is located at the start of spring in the Northern Hemisphere. This point, through which we draw the 0h line of right ascension—the prime meridian of east-west measurement in the heavens—is still sometimes known as the "First Point of Aries." This is because Aries was the constellation that the vernal equinox was in about two thousand years ago when the point was named. It is the precession of Earth's axis which is responsible for this movement of the equinoxes. It continues and will bring the vernal equinox into Aquarius—beginning the true astronomical "Age of Aquarius"—in the year 2597.

Pisces is hardly any brighter than Piscis Austrinus—without Fomalhaut. Within the bounds of this rambling constellation are two magnitude 3 1/2 stars and a lot of 4th- and 5th-magnitude ones. One of the unexpected pleasures of having a crystal-clear, moonless night in the country in early fall is being able to look up and trace the dim lines of Pisces easily with the naked eye. For most of us these days, however, binoculars are in order.

The easiest section of Pisces to identify is the westernmost part. Here again the Great Square of Pegasus helps. Look about one height of the Great Square below (south of) the Great Square to find a rough circle of faint stars less than a quarter the size of the Square. This is the Circlet, representing the head of the western fish of Pisces. It lies about an hour of right ascension east of, but a little farther north than, the Water Jar asterism of Aquarius. Just off the east edge of the Circlet is one of the sky's reddest naked-eye stars—though you'll need a small telescope to detect the color. This is 19 Piscium, better known by its variable star name, TX Piscium. It usually shines at between 4.9 and 5.1, though its extreme range is 4.5 to 5.3.

The overall pattern of Pisces is like a giant checkmark with arms opening around the southeast corner of the Great Square. The arms of the checkmark are the two fish, one pointing west (its head being the Circlet) and one pointing north or slightly northwest. The tails of the two fish are tied together with a cord. This, the point of the checkmark, the easternmost part of Pisces, is marked by the magnitude 3.8 Alpha Piscium—also known as Risha or Alrescha. The name comes from the Arabic for "the rope" and it does indeed mark the knot in the cord which binds together the two tails of the fish. But actually "Risha" refers to a rope in another early Arab imagining of the constellations in this part of the sky—an imagining in which parts of Pisces and Andromeda were a rope and the Great Square was a bucket.

The northern head of Pisces, up near Andromeda, is incredibly faint. The last reasonably bright star marking northwest along the body of the northern fish is magnitude 3.6 Eta Piscium (brightest star in Pisces). And just 1 1/2° east-northeast of this star is one of the most elusive of all the Messier objects, the galaxy M74. This 9th-magnitude galaxy is spread out over a diameter of about 10', so it is important to have a dark sky and low power when looking for it. M74 is the first object that observers must see, low in the dusk on a March night, if they want to behold all the Messier objects in one night. (For more on these Messier marathons, see the March chapter.)

Pisces does have a few pretty double stars to offer. Alpha Piscium is one itself, though rather tight—a 4.2 and a 5.1 star 1.8" apart and very, very slowly closing (they will reach the closest together they appear from Earth in the entirety of their 720-year orbit—1.0" apart—in 2074). There is controversy about the colors of the components of Alpha Piscium. Take a look and see what you think they are.

Alpha Piscium—Risha—has received some additional fame from a role it plays for stargazers interested in observing another celestial object—the brightest of all the long-period variable stars of large range. That variable star is one of the special attractions of our next constellation, Cetus the Whale.

CETUS (AND THE SCULPTOR GALAXIES)

Cetus the Whale is a fittingly large constellation and it is rather bright on either end—but not in between. It also travels across the heavens tailfirst, so we might as well note first that this tail is marked by the

Whale's brightest star, magnitude 2.0 Deneb Kaitos ("tail of the whale"). Deneb Kaitos is also called Diphda—from the Arabic for "the frog." For it was in an early Arab imagining "the first frog" and Fomalhaut "the second frog." As you gaze low across the south and southeast sky on October evenings, you can see how that vast dim expanse is indeed broken by only two really bright stars, Diphda and Fomalhaut. Considering them frogs is consistent with the idea of this whole region of the heavens being the Water or Great Celestial Sea— though it may seem a bit undignified!

The Sculptor Galaxies South of Cetus

Beta Ceti is useful as a reference point for getting to some wonderful, though rather southerly galaxies. The most outstanding of these galaxies well south of Beta Ceti is NGC 253. Except for M31, the Great Galaxy in Andromeda, this is the finest-looking galaxy in autumn skies—that is, if you don't live too far north and if you observe when NGC 253 is at its highest in a dark sky. NGC 253 is a splendid spiral shining at magnitude 7.1 and measuring a spectacular 25' by 7'. Mottling caused by dust features starts becoming prominent in it with 8-inch and 10-inch telescopes in good sky conditions.

NGC 253 actually lies just over the boundary from Cetus in the dim and southerly constellation Sculptor. NGC 253 lies at –25° declination so is not too far south to be hopelessly dimmed by haze and ordinary atmospheric absorption. Unfortunately, the second-best galaxy in Sculptor is a different story, for it lies on the southern border of the constellation at declination –39°. This is NGC 55, which glows at magnitude 7.9 and measures 32' by 6'. It lies only about 7 million light-years from Earth, NGC 253 about 7 1/2 million.

The Head of Cetus

Two hours of right ascension east of Beta Ceti is the head of the Whale. In the middle of October evenings, observers at midnorthern latitudes will find it at about the same height in the east as Beta Ceti is in the southeast. But later in the night (or on November evenings), the head rises higher and we see that it is actually well to the northeast of the tail. As a matter of fact, the head of the Whale sticks up just above the celestial equator into the northern celestial hemisphere. The head is a fairly compact, roughly circular pattern of stars which seems to be on a long stalk of a neck—very un-whalelike. Actually, the tail of this constellation looks more like a whale's nose and the head more like a whale's tail! There may be a good explanation for why the head looks so unlike a whale's, however. Cetus is Latin for "whale," but these stars are really known in myth as a sea monster of unspecified kind—a sea monster which was coming to devour the maiden Andromeda and had to be killed by the hero Perseus. (We discussed this wonderful myth in the December chapter near the start of this book.)

Whatever it does or does not look like, the head—and neck—of Cetus is interesting. The brightest star of the head, though normally only the second-brightest of Cetus as a whole, is magnitude 2.5 Alpha Ceti, also known as Menkar. The bottom of the head is magnitude 3.5 Gamma Ceti (Kaffaljidhm). It is a rather tight but attractive double star—a magnitude 3.5 primary with a 6.2 secondary 3″ from it.

The naked-eye star just south of Gamma Ceti could be considered the start of the neck of Cetus. It is magnitude 4.1 Delta Ceti and it lies only 0.5° north of the celestial equator. More importantly for observers of deep-sky objects, Delta Ceti is only 1° northwest of the galaxy M77.

Although the total brightness of M77 is only 8.9, it is a completely different sight than the elsuive M74 in Pisces. M77 is the apparently brightest example of a Seyfert galaxy, a type of galaxy with an intense, brilliant, energetic core. The Seyferts are a puzzle, but they seem to be cousins of the quasars, though far less powerful. The nearest quasar, however, is a 13th- or 14th-magnitude object about 2 1/2 billion light-years away (3C273 in Virgo—see the May chapter). The starlike core of M77 shines at about 10th magnitude from perhaps as close as 47 million light-years from Earth.

Mira, the Wonder of the Whale

Most of the time, the long gap between Delta Ceti and the first star of the whale's body (Zeta Ceti) is devoid of any star that can be easily glimpsed with the naked eye. But for a few months out of every eleven months another naked-eye star appears right in the line of the neck (about one diameter of Cetus's head southwest of the head). For at least a number of weeks that star is brighter than the competition a few degrees to its east-northeast—magnitude 5.5 star 66 Ceti. Most years, it

MIRA

An Idealized Light Curve

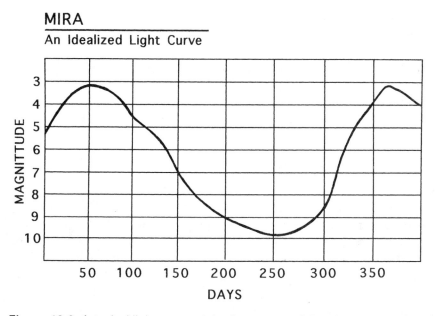
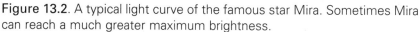

Figure 13.2. A typical light curve of the famous star Mira. Sometimes Mira can reach a much greater maximum brightness.

grows as bright as magnitude 4.0 or 3.0 and some years even brighter. About once a decade or so, it may outshine magnitude 2.5 Alpha Ceti and even rival magnitude 2.0 Beta Ceti. Once, in 1779, it reached almost first magnitude, becoming nearly the equal of Aldebaran! Yet at minimum light this star is only ninth magnitude or dimmer. Its Greek-letter designation is Omicron Ceti. But it is known as Mira Ceti, "the wonder of the Whale," or most often just Mira, "the wonderful."

Mira is the brightest and prototype of the long-period pulsating variable stars. Mira itself rivals Betelgeuse as the largest star known. It is probably capable of filling our solar system out to about the orbit of Jupiter! The star is so cool that ordinary steam exists in it. Its period is fairly regular, averaging about 332 days. That's about eleven months and, consequently, the peak brightness occurs about one month earlier in the calendar each year. For a few years in every twelve, then, the peak is hard to observe because it occurs in the late spring and early summer months when the Sun is in the same general region of the heavens as Mira. Unfortunately for readers of this book in the first year or two of the book's availability, these are the years when Mira's peak is hidden. The situation improves greatly by 2006.

The discovery of Mira is credited to the Dutch astronomer David Fabricius, who recorded it on August 13, 1596. But he seems to have thought the star was a nova. Since the following years were ones when the Sun hid the peak, the true nature of the star was not realized until later. The next time Mira was seen was in 1603, when Bayer, not realizing its true nature, observed it as a 4th-magnitude star and cataloged it as Omicron Ceti.

THE ORIONID METEORS

In the May chapter, we discussed one of the annual showers of meteors which are derived from Halley's comet. Those Eta Aquarid meteors, we noted, come from a radiant which is quite low when morning twilight arrives for observers at midnorthern latitudes. The observing situation is much better for the second of the Halley showers, the one which is at its best for a few days around October 20–21. This shower's radiant is at its highest, very high in the south around 4 A.M.—well before October nights are over. So if you live in the Northern Hemisphere and want to see flaming streaks caused by the dust-tail debris of past visits of Halley's comet, by far your best chance is this month.

The October meteors from Halley are known as the Orionids. Their radiant is actually located near the northern bounds of Orion the Hunter, up in the club of faint stars he extends to the northeast from his bright body. This spot is up near the feet of Gemini the Twins.

Like the Eta Aquarid meteor stream, that of the Orionids is a complex one, consisting of various strands of meteoroids. So the mornings when you will see the most is slightly variable from year to year for this astronomical reason—also for your local meteorological ones and the phase of the Moon, too, of course. In a typical year if you are diligent and willing to go outside so late (or early) you might be lucky enough to see ten or fifteen Orionids in your best hour, away from city lights. But a few times over the years, you may get to see several times as many Orionids per hour.

OCTOBER'S OBSERVING EXPERIENCE

My predawn encounters with the Orionids have brought me some of my most moving observing experiences. When I was a child, I remember going out around 4 A.M. and being awed by the vast stillness and majesty of the stars on that breathtakingly clear and dark night. Another morning before dawn, in 1985, I was looking at Halley's comet in a small telescope when it was located near the radiant of the Orionids—and when I would look up from the eyepiece I caught sight of Orionids! Even though these meteors must have originally been from previous visits of the comet, this was a once-in-a-lifetime experience of the meteors and their parent object together.

Several times, when I have watched for Orionids or other stellar sights on an October night, I have seen morning twilight brighten to slowly reveal the glowing hues of yellow, orange, and red leaves all around me. Whether you observe before dawn or on an evening in October, it is remarkable to think of that blaze of colors surrounding you but hidden by the cloak of night—just as the wonders of the stars are hidden by the cloak of day. In either case, a miraculous world of beauty is close at hand (even the stars seem close on a clear night), ready to be revealed to our unknowing selves with what seems like the touch of a magician's wand.

In those still October nights across much of the United States, the southing geese are still heard. They are spurred on their migration by colder weather up north. But in October that cold weather starts arriving to those of us who live south of Canada. Heavier coats may be needed as the night progresses. The first killing frost of the year occurs around the time of the Orionids, or a bit later, for many locations around 40° N. It's remarkable to think of that adornment of crystalline ice feathers forming with silent patience as you stand out watching the stars on an October night (hope it doesn't coat your optics, though!).

October ends with Halloween and some years the Full Moon after Harvest Moon—the Hunter's Moon—may occur late in the month and it will rise in dusk . . . maybe like a pumpkin, in this case made orange by some horizon haze during a spell of Indian summer (a warm period after the first frost). There is a lovely group of stars which has an intimate connection with Halloween, but that story is told, with the rest of the legends about the Pleiades, in the December chapter.

Figure 14.1. Though discussed in the December chapter, the V-shaped Hyades star cluster (lower left) and tiny dipper shape of the Pleiades (near center) are already visible on November evenings. The bright dot of Saturn is at lower right in this photograph by Steve Albers.

Chapter 14

November

[Constellations covered: Andromeda; Aries, Triangulum; Cassiopeia; Camelopardalis]

I n last month's chapter, we examined mostly the dim constellations of the Water or Great Celestial Sea, those star patterns which fill much of the southern sky on autumn evenings. Now, in November, we take a more northerly and much brighter route through the heavens. High in the north and northeast, trailing our one single bright constellation from last month—Pegasus—are three constellations. They have stars as bright as those of Pegasus but in more compact and therefore more prominent form. The three are Andromeda the Chained Maiden, Cassiopeia the Queen, and Perseus the Champion (or Hero). They are connected with last month's Pegasus the Flying Horse and Cetus the Whale, and with September's Cepheus the King, in one great Greek myth which brings together these six neighboring constellations of the autumn heavens. In this chapter, we will examine Cassiopeia and Andromeda. Perseus and the great myth were discussed in the December chapter which led off our year of the stars, earlier in the book.

Our search of the high north and high east sky in November also takes us to small but fairly bright star patterns twinkling near Andromeda. They are the patterns of constellations Aries the Ram (a famous constellation of the zodiac) and Triangulum the Triangle (scarcely known at all to people outside astronomy). Another pattern is lower in the east than Aries and Triangulum and is treated fully in our December chapter. But so eye-catching is this pattern that we must at least acknowledge it this month. It is the tiny dipper-shape of the loveliest of all naked-eye star clusters, the Pleiades. (See figure 14.1.)

We may feel that we really need a lot of brightnesses and wonders packed into clear November nights for a good reason: There aren't many clear November nights. Nor'easters begin to plague the Atlantic coast; windstorms and lake-effect snowstorms trouble the lands around the Great Lakes. There's no denying that for most mid-northern lands November brings a dramatic increase in the amount of storminess and cloudiness after the clear period of September and October.

The poet Thomas Hood wrote: "No shade, no shine, no butter-flies,/No fruits, no flowers, no leaves, no birds,—No-vember!" But no one can say that this month's clear evenings contain no bright stars and other stellar wonders. Within the first of our featured constella-tions for November we will meet the Milky Way's big sister—whose light is like a "candle seen through horn." The constellation is Andromeda, and the object is M31, the Great Galaxy in Andromeda.

ANDROMEDA

Andromeda: the very name conjures up adventure and excitement! That is true whether you are applying the name to myth's first great damsel in distress, the first constellation in alphabetical order and right ascension, or the favorite and first of all sights beyond our galaxy that skywatchers look for. Andromeda is all three things: a mythic maiden, a constellation named for the maiden, and a great galaxy which lies within that constellation.

First, let's look at the constellation, and the stars which compose it.

The Stars of Andromeda

As we noted in the previous chapter, one of the main stars of Andromeda—Alpha Andromedae—is used to form the northeast corner of the Great Square of Pegasus. Thus the pattern of Andromeda is liter-ally connected to that of Pegasus. What is the pattern of the constellation Andromeda? Whatever other details you may add to it, the most basic part of the pattern is a slightly curved line radiating northeast away from Alpha Andromedae. And that line is remarkable for several reasons.

The first reason is that the three stars in it are of essentially the same, rather impressive brightness. Alpha Andromedae is brighter than the other stars of the Great Square of Pegasus. But it is virtually identical in brightness—magnitude 2.1—with the other two stars in the line of Andromeda. The three stars are, in order, southwest to northeast: Alpha, Beta, and Gamma Andromedae. Their proper names are Alpheratz or Sirrah (Alpha), Mirach (Beta), and Almak (Gamma). And they are separated from each other by almost exactly the same amount. It happens to be almost exactly one hour of right ascension from Alpheratz to Mirach and the same distance from Mirach to Almak. What's more, the (epoch 2000) positions of each is similarly near the start of each one's particular hour: Alpheratz is at 0h08m, Mirach at 1h10m, and Almak at 2h04m.

In other respects, the three stars offer a neat proportional scale, and some marvelous contrasts. For instance, from Alpha to Beta to Gamma the luminosities and distances increase, (almost) proportionally: Alpheratz is 98 light-years from Earth, Mirach 200 light-years from Earth, and Almak 370 light-years from Earth. So each is roughly twice as far and four times as bright as the star before it. Their colors and spectral classes come in a pleasing range. Alpheratz is a B9 blue-white star. Mirach is an M0 "red" (deep orange-yellow) star. And Almak? With binoculars or a very small telescope at low power, it should look orange-yellow, though much lighter in hue than Mirach, for Almak is a K3 star. Or, rather, we should say, this is what the color of the primary component of Almak is.

Almak—Gamma Andromedae—is one of the finest double stars in the heavens. The primary is a 2.3-magnitude star with a 5.4-magnitude B9 companion 10″ away from it. The colors rival those of the great double star of Cygnus, Albireo. They have been variously described as topaz and aquamarine, orange and emerald, or just yellow and blue. The pair is considerably tighter than Albireo, so somewhat higher magnification is called for. Interestingly, the dimmer B star is itself a spectacular double—but only if you have a large amateur telescope and superb atmospheric steadiness when the C star (magnitude 6.0 or 6.3, according to different sources) is at about its widest separation from B. Even the widest separation of B from C is only 0.6″ and in 2001 it was 0.4″ decreasing toward a minimum of 0.1″ in 2013 (the period is sixty-one years, so B-C won't be much easier to split anytime soon). Most of us will have to be content simply to know of the existence of this pair. The A-B duo is, however, a sight which everyone with a working astronomical telescope can—and should—observe.

About 5° south-southwest of Gamma is a huge marvelous open cluster with a striking wide double star near its edge. The cluster is NGC 752. It shines at magnitude 5.7 and measures 50′. In a rich-field telescope, about sixty stars are seen splashed here. And accompanying the superb cluster is 56 Andromedae: a pair of almost equally bright (5.7 and 5.9) stars 201″ apart, both with orange hue. The slightly brighter star is a K0 sun with a B-V color index of 1.1 and the slightly fainter star is an M0 sun with a color index of 1.6. Do you notice this difference visually? These two stars are at much different distances. The brighter is about 320 light-years, the dimmer about 900 light-years from Earth. And the cluster NGC 752 is believed to lie roughly 1,500 light-years from us.

As interesting as the bright stars of Andromeda are, they are often neglected because everyone wants to see the constellation's—and northern hemipshere's—premiere galaxy—M31, the Great Galaxy in Andromeda.

The Great Andromeda Galaxy

To locate this object, go from Alpha to Beta Andromedae and then "make a right"—a right turn a few degrees to magnitude 3.9 star Mu Andromedae. Another few degrees brings you to magnitude 4.5 Nu Andromedae, but if your sky is reasonably dark you will already see, with the naked eye, an amazing object looming just past Nu. It looks like an elongated smudge of light, its long axis oriented almost parallel to the long Alpha-Beta-Gamma line and therefore perpendicular to the Beta-Mu-Nu line. Under fairly good conditions, binoculars or naked eye might perceive the smear of light stretching for 1 1/2° or 2°. But under truly excellent sky conditions the naked eye can glimpse M31 extending for an awesome 4° or even 5°.

When we see with the unaided eye this slip of seeming phosphorescence, as fanciful and delicate as a little toy hanging on a mobile over a baby's crib, what are we actually seeing? A pinwheel of approximately 400 billion (400,000,000,000) suns. It takes the combined radiance of them all to produce this magnitude 3.4 glow from a distance of more than 2 million light-years. M31 is the most distant object that can be seen with the naked eye in less than a superbly dark sky.

The Great Galaxy in Andromeda is a spiral galaxy like the Milky Way except that it is our galaxy's big sister. M31 is estimated to be about one and a half times as wide and twice as massive as the Milky Way (M31 is about 300 billion times as massive as the Sun), and possesses about twice as many stars. Big sister Andromeda has a slightly tighter spiral structure, two dust arms, and five outer arms of star clouds. It seems to be less active in star formation than the Milky Way but it has about six hundred globular star clusters (about three times as many as the Milky Way), and may possess not just one but two massive black holes at its center.

Most of the light reaching us from M31 is from the central part of it. Even on a night when the naked eye can trace the galaxy's length out to 4° or 5°, the magnification of a typical telescope will spread out the already dim outer extensions of the galaxy so much that they cannot be detected through the eyepiece. This leads to disappointment for many first-time telescopic observers of M31, who in any case expect to see something very much like the stunning long-exposure photographs of the galaxy. We should also remember that M31 is tilted just 13° from edge-on, so that its spiral arms and dust lanes are largely superimposed on one another as seen from our vantage point. (How spectacular M31 would be if we could see it from an "overhead" or "plan" view as we do with the galaxy M51!)

In truth, however, the telescopic view of M31 *is* spectacular. It's just that you need to learn to observe carefully and to understand what you are seeing. You also need to combine in your mind the vision of what you are seeing of M31 through the telescope with the very different view through binoculars and a third different view of it, the one with the naked eye. No celestial object has grandeur that is so crucially the sum of the different views of it with different instruments.

How much detail in M31 can you make out with a 6-inch or 8-inch telescope? You can start to glimpse a few of the dust lanes that are so prominent in photographs (the most prominent part of one lies northwest of the galaxy's center). You can make out a patch slightly brighter than its surroundings in the outer parts of the telescopic M31 southwest of the galaxy's center: this is the galaxy's bright star cloud NGC 206. What about in the middle of M31? The center is usually a slightly yellow hazy ball a few arc-minutes wide. But on a night of good "seeing," you can note at high power (200× or more) that there is a star like "nucleus"—a 2.5" by 1.5" intense concentration of blended stars at the galaxy's center. This may consist of roughly 10 million stars in an area only about 50 light-years across!

Actually, at low magnification with a medium-size, or even a small telescope, the details that an observer may well notice first are not in M31 itself. The observer may notice one or both of the close satellite galaxies: M32 and M110.

You might detect M110 first because this definitely elongated elliptical galaxy is larger and farther removed from M31—about 0.6° northwest of the center of M31. M32 is of about the same brightness—magnitude 8.1—but is much closer to the glow of M31 and is small and roundish, even capable of being mistaken for a fuzzy star at low power with a subpar telescope or in poor "seeing."

By the way, M110 was originally (and still sometimes is) called NGC 205. The final version of the catalog which Messier published in his lifetime ended with M103. But additional objects which he saw (and in the case of M110, discovered) have been added over the years and these past few decades most astronomers seem to have accepted the 110-object list.

M32 is one of the smallest galaxies known, only about 6,000 light-years wide. It has "only" about 2 billion stars. It is also thought to have a black hole in its center. Interestingly, M110 may be getting rejuvenated into a period of more active star formation by the pull of M31.

M32 and M110 are the closest companion galaxies of M31, but astronomers know a number of others. The brightest of these, NGC 147 and 185, are companions of M31 about 12th-magnitude and a full 7° north of M31, over the border in Cassiopeia.

M31 and the Milky Way are the dominant objects in the Local Group of galaxies. They are opposite ends of the Local Group. All in all, M31's family contains at least thirteen galaxies, the Milky Way's at least twelve. There are some independent galaxies in the Local Group, too, so the grand total for the number of members of the Local Group is more than thirty and continues to increase as the elusive traces of still other small galaxies are identified.

M31 must have been noticed in ancient times but we have no records to prove that. According to astronomer R. H. Allen, "it is said to have been known as far back as A.D. 905; was described by [Persian astronomical writer] Abd al-Rahman al-Sufi as the Little Cloud before 986."[15] The first telescopic observer of it appears to have been Galileo's rival Simon Marius, who first viewed it on December 15, 1612. It is often published that Marius compared the telescopic sight of M31 to "the light of a candle seen through horn" but a fuller translation, according to space and astronomy author Willy Ley, is "a candle flame seen through the horn window of a lanthorn [lantern])."[16] That's a beautiful and accurate description of the hazy and therefore seemingly translucent central hub of M31 as seen in a small telescope.

For hundreds of years after Marius, M31 was just a particularly impressive and odd example of a nebula—a hazy patch of light which could not be resolved into individual stars in the telescopes of previous centuries. It was called the "the Andromeda Nebula" and indeed that old name was sometimes used even in books of the midtwentieth century by popular writers decades after scientists knew better. When an exploding star appeared in the "Andromeda Nebula" in August 1885 astronomers still were not aware of what M31 was or how far away it was and supposed the star was a less-than-catastrophic form of star explosion, a nova. This star—dubbed S Andromedae—brightened to reach about magnitude 5 1/2 before fading away. Only when the true nature and distance of M31 was discovered did astronomers realize that S Andromedae was, briefly, the farthest star ever bright enough to see with the naked eye. It had reached a luminosity about a billion times that of the Sun and was a member of a new class of exploding star, a "supernova."

The discovery of the nature of galaxies and of the much vaster scale of the universe were derived from the identification of Cepheid variable stars on photographs of M31 made by the 100-inch telescope at the Mt. Wilson Observatory in California. The luminosity of Cepheids is related proportionally to the length of their periods, enabling their distance to be determined. In 1924 Edwin Hubble announced that the distances to the Cepheids in M31 was about 900,000 light-years. Later it was learned that there were two types of Cepheids and that the correct distance was more like three times farther. There is still debate over the exact distance of M31 but after a proposed increase in distance in the 1990s, the experts seem to have settled back on 2.2 or 2.3 million light-years. But the revelation which came from Hubble's study of M31 was monumental: "extragalactic nebulae"—now called galaxies—were systems of the same general kind as our Milky Way, at distances enormously greater than previously supposed. The Milky Way was no longer the center of the universe and that universe was immensely larger than scientists had guessed.

The fate of the two major galaxies of the Local Group—M31 and the Milky Way—is now thought to be known, and it couldn't be more startling: They will collide. This should happen in about 4 billion years. The Andromeda Galaxy will loom gigantic and then perhaps sift right through the Milky Way. There is even a small chance that the collision could eject our solar system—without any noticeable jostling of the planets—into the depths of intergalactic space. This would all happen about a billion years or so before the Sun becomes a red giant and swells out to melt Earth's surface or engulf our world—or so researchers thought just a few years ago. The latest thinking is that the Sun may become a red giant much sooner, Earth's oceans be boiled away and our planet rendered inhabitable even sooner (a mere billion years from now). And a recent study suggests that if the human race has moved farther out in our solar system or to another solar system to stay cool, it may see the Andromeda Galaxy arrive and *not* sift through but instead merge permanently with our own. The process would no doubt be accompanied by tremendous fireworks of supernovas and starbirths and other disruptions but might be quite survivable by most life-forms in the two galaxies.

You needn't alter any of your plans for next week because of a galaxy collision that is due in 4 billion years. But when you watch the Andromeda Galaxy tonight, pause to reflect that it is getting closer to you, to all of us and our galaxy—fifty miles closer every second.

ARIES AND TRIANGULUM

The little patterns almost exactly due south of Gamma Andromedae are those of Aries the Ram and Triangulum the Triangle. Aries, as a constellation of the zodiac, is tremendously more famous than Triangulum. But the latter contains an important link to M31, so let's explore it first.

M33 and Triangulum

The pattern of Triangulum is a small and skinny triangle of modestly bright stars. Alpha, Beta, and Gamma Trianguli shine at magnitudes 3.4, 3.0, and 4.0, respectively. Interesting double stars in the constellation include Iota Trianguli (magnitudes 5.2, 6.7; separation 3.9″; yellow and blue) and 15 Trianguli (5.4, 6.7; 142″; orange and blue). The outstanding observational target in Triangulum, however, is M33.

M33 is a spiral galaxy thought to be only a little more distant than M31. But, unlike M31, it is presented to us nearly face-on. We should therefore have an ideal view of its spiral arms curling out from around its center. In professional images, M33 does appear as a showpiece worthy of the nickname sometimes given to it: the Pinwheel Galaxy. Yet this object has a long reputation of frustrating would-be observers. The famous story about M33 is of observers being unable to detect this magnitude 5.7 galaxy even with a 10-inch telescope—yet seeing it in binoculars!

As you might expect, the explanation for this story is that the galaxy's brightness is spread over a large area. Not as large an area as M31, to be sure, but M33 is far less bright. The Triangulum Galaxy (another nickname for M33) measures 71′ by 42′ on photographs and although almost all of its light comes from an area about half that size, this is still larger than the apparent size of the Moon. Any substantial light pollution will therefore render this giant low-surface-brightness object invisible—as will using too high a magnification even in moderately dark skies. The key is to try low magnifications on dark and clear nights, and to look at first for something like a big soft blur. At first that's all M33 will appear to be. But under good conditions the eye can teach itself to see more and more of this galaxy, glimpsing the central region and tracing a giant loose reversed letter S of spiral arms by the brighter splotches of star clouds and nebulas here and there along them. The brightest feature in M33 other than its small center is one of the biggest, most luminous nebulas known— NGC 604. NGC 604 lies 12′ northeast of M33's center and only 1′ northwest of a magnitude 10.5 star.

M33 is similar in distance to M31, or perhaps a little farther. Thus it may be the most distant object really visible to the naked eye by any but the most skilled and owl-eyed observer in virtually pristine dark skies. As a matter of fact, the naked-eye visibility of M33 was a popular test of dark skies for the late deep-sky master, Walter Scott Houston, and also has been for a long time for the great comet observer John Bortle. I find that M33 can be glimpsed with the naked eye using averted vision when it is fairly high if stars of about magnitude 6.5 are visible. Such skies are becoming more and more uncommon in these days of widespread major light pollution, but if you get even better conditions M33 can actually become a reasonably prominent naked-eye sight, standing out as a patch of glow even if you didn't know exactly where to look for it.

Of course, in skies where M33 has borderline visibility, you do have to know its exact position. Even if you are trying to find it with binoculars or telescope, you should use some handy guide objects. One interesting fact is that M31, Beta Andromedae, and M33 lie in an almost straight line, with M33 not much closer to the southeast of Beta than M31 is to the northwest of the star. M33 can also be located with reference to the point of the triangle pattern of Triangulum, Alpha Trianguli. M33 lies 4° west-northwest of Alpha Trianguli (which is sometimes called Mothallah from the Arabic for "the triangle").

M33 could be considered the third major galaxy of the Local Group, but it has only about half the width, one-fifth the luminosity, and one-seventh the mass of the Milky Way, and is even less impressive in these properites compared to M31. As a matter of fact, M33 is a member of M31's family and lies only 700,000 light-years from M31—so the Andromeda Galaxy would be a very much more impressive sight in the sky of beings living in M33 than it is from Earth or anywhere else in the Milky Way. On the other hand, it is true that M33 is more active with starbirths in the clumpy glows of its spiral arms than the Milky Way, and much more active than M31.

Aries the Ram

Just south of Triangulum is a constellation with an even smaller main pattern. Aries the Ram also uses just three stars for its main pattern, which looks more like a bone than a ram. Yet among the three is one of the autumn constellations' brightest stars, magnitude 2.0 Alpha Arietis, also known as Hamal. Hamal (from the Arabic for "the ram") is distinctly orangish (in binoculars or telescope) and lies just 66 light-years from Earth. Interestingly, magnitude 2.6 Beta Arietis (Sheratan—from an Arabic term meaning "the two signs," in this case Aries and Pisces) is located at a similar distance, 59 light-years from us. Gamma Arietis (Mesarthim—from the Arabic for "the fat ram") shines at a lowly magnitude of 4.6 but is a spectacular double star. The two components are both magnitude 4.8, their separation 7.5″. Their spectral classes are B9 and A1, so that most observers see them as white. Interestingly, the position angle of the second star in relation to the first, which is hardly changing, is 0°—that is, exactly due north. The duplicity of Gamma Arietis was discovered as early as 1664, by Robert Hooke when he was hunting for a comet.

Although Aries contains a few additional pretty double stars, it is remarkably poor in other deep-sky objects. Aries and Libra are the only zodiac constellations which contain no Messier objects.

The great interest of Aries remains its role as a zodiac constellation. As discussed in last month's chapter under Pisces, the vernal equinox point is still known as "the First Point of Aries," for it was located within the Ram about two thousand years ago.

CASSIOPEIA

Our next constellation, Cassiopeia the Queen, is the northern neighbor of Andromeda. Whereas Andromeda appears pretty much overhead in the middle of November evenings at midnorthern latitudes, Cassiopeia is truly a north circumpolar constellation for those latitudes (about 30° N to 50° N). That means you will have to turn and face north to observe it.

The main pattern of Cassiopeia is second in prominence only to the Big Dipper in the north circumpolar region, and lies in almost the opposite direction from the pole in relation to the Big Dipper. Thus, when the Big Dipper is high above the North Star—as on spring evenings—Cassiopeia is far below the North Star (and hard to see down by the north horizon, perhaps hidden by trees—or by haze or light pollution, which are at worst low in the sky). But on autumn evenings, when the Big Dipper is low and hidden and thus can't be used for finding the North Star, Cassiopeia is high and useful.

It's appropriate that Cassiopeia is near Andromeda, for in myth they are mother and daughter. The story of them and Perseus is given in the December chapter.

The Stars of Cassiopeia

The main pattern of Cassiopeia is a sort of zigzag of second- and third-magnitude stars (though in the lore of the Elves of J. R. R. Tolkien's fantasy works it was Wilwarin, the Butterfly). The zigzag looks somewhat like a capital letter M on fall evenings when it is high in the north. The pattern's three brightest stars are those which lead it—the most westerly of the pattern in right ascension and, on fall evenings, also the most westerly in relation to the cardinal points of the sky. West to east, the three stars are Beta, Alpha, and Gamma Cassiopeiae. Their magnitudes are 2.3, 2.2, and between 2.2 and 2.5, respectively.

Gamma Cassiopeiae is a quite unusual type of variable star. It is a rapidly rotating B star which shows a special spectral signature when it gets active, probably expelling matter. Gamma had burned at about magnitude 2.25 until the period 1910–1936, during which it brightened ever-so-slowly by half a magnitude. By April 1937 it brightened a bit more—to a maximum of 1.6—but by year's end dimmed all the way back to 2.25. A slow further dimming, to 3.0, occurred in the next three years—then a very slowly brightening back up to its current range (2.2 to 2.5). No star of this rare type so bright in our skies was identified until Delta Scorpii performed somewhat similarly (though much more rapidly) around the turn of the twenty-first century (see July chapter for details). Both Gamma Cassiopeiae and Delta Scorpii certainly bear continued watching.

The final stars of the Cassiopeia "M" are Delta (magnitude 2.7) and considerably dimmer Epsilon (3.4). The proper names of Beta, Alpha, Gamma, Delta, and Epsilon Cassiopeiae are sometimes said to be Caph, Shedir, Tsih, Rukbah, and Sagin or Navi. Interfering with the pattern a bit is Eta Cassiopeiae, a star as bright as Epsilon, a bit off of and short of halfway along the line from Alpha to Gamma. Also, you can add to the M pattern of five stars by extending the line from Delta to Epsilon about its own length onward to reach Iota Cassiopeiae (magnitude 4.5).

Now these additional stars are good to know, for both are superb double stars. Eta consists of a 3.4 and a 7.5 star that are 13″ apart. The primary is yellow but the secondary is variously described as ruddy purple, garnet, lilac, red, or orange. This pair is located only 19.4 light-years from Earth and the primary is exceptionally similar to our own

Sun in luminosity, size, and spectral class. Iota Cassiopeiae is actually a tight triple star, the 4.6 primary having a 6.9 star 2.5″ from it and a 8.4 star 7″ from it. The colors have been described as yellow, lilac, and purple or blue.

By the way, whereas five of the stars of the Big Dipper are at nearly identical distances from Earth, and the remaining two not much farther away, the major stars of Cassiopeia are all at very different distances.

We cannot leave discussion of the stars of Cassiopeia without mentioning the great supernova which appeared in Cassiopeia in 1572 just a bit northwest of Kappa Cassiopeiae (Kappa is the 4th-magnitude star which forms a rough square with Beta, Alpha, and Gamma Cassiopeiae). The astronomer Tycho Brahe was not the first to see and study the supernova but he was the most famous and important. The star burned about as bright as Venus, and was visible even in broad daylight, before fading from view. Two of the other five possible supernovas of the past thousand years have occurred in Cassiopeia, though the ones in 1181 and 1680 were apparently dimmed greatly by interstellar dust.

There are two stars in Cassiopeia which are among the most luminous in our part of the galaxy, almost like permanent novas. A few degrees from Beta Cassiopeiae is Rho Cassiopeiae, usually about magnitude 4.4 to 4.6 but with extremes of 4.1 and 6.2 that may be seen to occur in 2003–2004. The star is estimated to lie about 6,000 light-years away and to have an absolute magnitude of –8. Then there is Phi Cassiopeiae, which is almost as luminous as Rho if it (Phi) is a true member of the bright Cassiopeia cluster NGC 457.

The Clusters of Cassiopeia

Cassiopeia is a wonderland of open clusters. While no single one of them rivals the very greatest open clusters in the heavens, their sheer number and variety is marvelous.

One of the finest of the Cassiopeia clusters is NGC 457, located just south of Delta Cassiopeiae. If Phi Cassiopeiae is not a member then the cluster is still a magnitude 6.4 gathering of about eighty stars in a 13′ space. Phi is an orange supergiant which shines at magnitude 4.8. It was once thought to lie as far as 9,000 light-years away. But even if Phi and NGC 457 are only about 5,000 light-years from Earth, the star has an absolute magnitude of about –7.

NGC 457 has two nicknames: the Owl Cluster and the E.T. Cluster. Both refer to the shape formed by the brightest stars in the cluster, which is like a figure with two big eyes (one of them is bright Phi) and long, mostly downward-extended wings or arms. "E.T." is a direct reference to the extraterrestrial in the famous Steven Spielberg movie of that name.

Two Messier objects glow in Cassiopeia and both are open clusters.

M52 is a very rich cluster of about one hundred stars. Its angular diameter is 13″. Its total apparent magnitude is 6.9, and on the clearest night I've ever experienced, I believe that I glimpsed it with my naked eye! The cluster lies only 1/2° south of ruddy 4.9-magnitude 4 Cassiopeiae. The star and cluster can be easily found by extending the line from Alpha to Beta Cassiopeiae one of its own lengths onward.

M103, last object on the original Messier list, is encountered as soon as you leave Delta Cassiopeiae on the way to Epsilon Cassiopeiae. It packs a magnitude of 7.4 into a diameter of 6′ and its brightest stars form a wedge or arrowhead shape.

Among other lovely Cassiopeia clusters are 7th-magnitude NGC 663 and NGC 7789. The clusters of Cassiopeia offer many hours of enchanted viewing.

CAMELOPARDALIS

The north circumpolar constellation which follows Cassiopeia the Queen across the sky is . . . a giraffe. Camelopardalis the Giraffe has the reputation of being one of the faintest of all constellations. Surprisingly, however, it offers a few deep-sky objects for binoculars and telescopes which should really not be missed. (By the way, for technical reasons, Camelopardalis appears on the winter, not the autumn, map in the back pocket of this book.)

Two of the three brightest stars in the Giraffe are Alpha and Beta Camelopardalis, which shine at just 4.3 and 4.0, respectively. They are located almost directly east of the main pattern of Cassiopeia and almost directly north of the zero-magnitude star Capella—but far from both. What is interesting, however, is that Alpha Camelopardalis is estimated to be 4,000 light-years from us and therefore one of the most luminous stars in our section of the galaxy. It is a member of an

O-B association called Camelopardalis OB1 from which it is a "runaway star." The second-brightest star in the Giraffe, at magnitude 4.2, doesn't have a Greek letter designation or apparently even a Flamsteed number—it is known as Struve 385. And it may rival or surpass Alpha in distance and luminosity.

Now get out your binoculars. If you have been able to locate Alpha and Beta Camelopardalis, then scan to a point east of Beta, which would form an equilateral triangle with Beta and Alpha. You should find there an open cluster which actually is bright enough for me to have once "discovered" with the naked eye—magnitude 5.3 NGC 1502. This small (8' wide) circular cluster has a star as bright as 7th-magnitude and has been called "the Golden Harp" cluster. It lies about 3,100 light-years from Earth. Once you have NGC 1502 located, then note in large binoculars or a small rich-field telescope that there is a remarkable string of many similarly bright (9th- and 10th-magnitude) stars which extends northwestward from NGC 1502. Or perhaps we should say that it tumbles like a waterfall, at least 2 1/2°, southeast down to the cluster. This asterism is, after all, called Kemble's Cascade. It was named by Walter Scott Houston for the man who discovered it in 1980, Canadian amateur astronomer Lucian J. Kemble.

Last but not least in dim Camelopardalis is the brightest galaxy in the north celestial hemisphere to not have a Messier number—NGC 2403. This spiral galaxy, so beautiful in photographs, can be somewhat tricky to observe because it is presented nearly face-on to us and its magnitude 8.4 brightness is spread over dimensions as great as 18' by 11' (though the galaxy looks somewhat smaller visually). Look for it about 5° northwest of Omicron Ursae Majoris (Muscida), the star which marks the nose of Ursa Major, and just 1° west of sixth-magnitude 51 Camelopardalis. In space, the galaxy is located about 10 million light-years from Earth.

THE LEONID METEORS AND NOVEMBER'S OBSERVING EXPERIENCE

In most years, the meteors which shoot very swiftly out from the Sickle of Leo the Lion around November 17–19 come in rates of only about five per hour in their best hours when the radiant is as high as it gets, just before dawn. This is the typical display of the Leonid meteor shower. However, for a few years around each of the three times a century that the parent comet of the shower passes, there is a chance of Earth encountering one of the dense swarms of Leonids. When that happens, the skies can seem to open up and pour "shooting stars." They can streak by at rates of a few hundred per hour or even at rates of one thousand per hour, the rate at which some authorities say a meteor shower becomes a meteor storm. Even stronger Leonid torrents have been seen—for instance, the one that briefly fell at a 125,000, and perhaps even 500,000, per hour rate in 1966, overwhelming the stars in the Southwest United States.

At the 1998 peak of the Leonids, "only" about one hundred meteors an hour showed up. But most of them were fireballs— meteors brighter than any planet or star—and there were even a number of shadowcasters (meteors so bright they cast prominent shadows in the landscape). I saw one shadowcaster flashing and flaring like lightning as it passed over me. Another left behind a glowing cloud of ionized gas which grew to become a ring three times wider than the Moon and then bigger and dimmer until fading from my naked-eye view . . . about eight minutes after it had first appeared! At the 1999 peak of the Leonids, Europe and the Middle East got to see storm numbers of Leonids and for the first time ever it was con- firmed (with videocams by amateurs!) that meteoroids were crashing into the lunar surface, producing bright flashes. In 2000, a very bright moon was in the sky but there was still a rather good and bright Leonid show. Then in 2001 large numbers of people in the United States and Canada and in the Far East and Australia got to see hours of hundreds of meteors per hour building to a thousand and more Leonids per hour (several thousand per hour in the Far East). Even as morning twilight became bright I was still seeing brilliant streaks falling like rain in bunches (four or five in a second or two around the setting Orion) down the sides of the sky. The Leonid display of 2002 occured with the Full Moon but in a few parts of the world numerous Leonids were seen.

Unfortunately, no storms of the Leonids are due for a long time after 2002. It might not be until the end of the twenty-first century that Earth passes though one of the dense swarms of the Leonids in space and sees another Leonid meteor storm. Still, you can never tell.

We can in any case still look for the few but often very bright

Figure 14.2. By midevening in November, the most splendid of all constellations, Orion the Hunter, is appearing in the east, heralding the approach of another year of the stars. (Photograph by Steve Albers.)

Taurid meteors in early November. But those of us who experienced some of the great Leonid displays will not forget their glory. Each year, we'll all feel almost as if the silence was audible and darkness visible from lack of Leonids.

Yet there is a moment on November evenings every year so majestic that it resounds like thunder, at least in the heart and soul. I speak of the moment—soon after about 8 P.M. in mid-November—when Orion the Hunter leans up suddenly above the mists on the eastern horizon. In Tolkien's *Lord of the Rings*, this is an event which the beautiful voices of a company of tall Elves greet with song and then joyous celebration. Orion's three-star Belt stands vertical as he rises, his form looming gigantic. His evening entrance onto the stage of the heavens signals that one year of the stars has ended but that another has just begun.

Afterword

What do you do after you finish reading this book?

You probably have given it a first read, all the way through, in far less than a year. If so, then you can continue using the book for the rest of your year of observing, going back to each month's chapter as that month is happening in the world. But what do you do after you have completed one year of observing, have taken the full tour of the heavens all the way around?

When you have completed not just *A Year of the Stars* the book but also a year of the stars in the sky itself, I do not think you will have to ask that question. You probably won't have seen all the sights listed in the book in your first run through. More importantly, those you did see you will be eager to see again now that you have the first year of observing experience under your belt and understand better—to some extent on a physical, intuitive level from months of practice—how to observe and what to look for. The second time around, you will be able to compare this season's nebula or globular cluster with those of all the seasons you've already seen. You will also have almost certainly sought out magazines, Web sites, and books other than this one to enrich your knowledge of your favorite objects and topics. Furthermore, in the course of a year, you will probably have developed a yearning for a second telescope—larger or perhaps merely different but with its own merits—and you might be preparing to purchase it. My advice is not to rush into such purchases (though I might as well tell the wind not to blow in a hurricane!). But when you are ready to get another telescope, there's no denying that it will bring new wonders to delight you.

Most fundamentally, of course, the fact is that the heavens are too rich to ever be exhausted by a true devotee of them. Surely, the more you observe—even of the same celestial objects—the more there is to see. And if to the infinite variety of the heavens you add the infinite variety of your observing environment, there can be no chance whatsoever of any dull repetition.

Every year of the stars is a return of celestial objects which have become your familiar friends. But like real friends there is always more to learn about them, always a new light to see them in or by. Which brings me to the final reason that you should never need fear a deadening sameness in any year of the stars: each year, you yourself will be different. Why? If for no other reason, because the heavens will have changed you.

Appendixes

APPENDIX 1: THE GREEK ALPHABET

α	alpha	ν	nu
β	beta	ξ	xi
γ	gamma	o	omicron
δ	delta	π	pi
ε	epsilon	ρ	rho
ζ	zeta	σ	sigma
η	eta	τ	tau
θ	theta	υ	upsilon
ι	iota	φ	phi
κ	kappa	χ	chi
λ	lambda	ψ	psi
μ	mu	ω	omega

APPENDIX 2: THE BRIGHTEST STARS

		Apparent Mag.	Distance (in light-years)
1.	Sirius (Alpha Canis Majoris)	−1.44	8.60
2.	Canopus (Alpha Carinae)	−0.62	313
3.	Alpha Centauri (Alpha Centauri)	−0.28c	4.40
4.	Arcturus (Alpha Boötis)	−0.05	37
5.	Vega (Alpha Lyrae)	0.03	25.3
6.	Capella (Alpha Aurigae)	0.08	42
7.	Rigel (Beta Orionis)	0.18	773
8.	Procyon (Alpha Canis Minoris)	0.40	11.4
9.	Achernar (Alpha Eridani)	0.45	144
10.	Betelgeuse (Alpha Orionis)	0.45v	522
11.	Hadar (Beta Centauri)	0.61	526
12.	Altair (Alpha Aquilae)	0.76	16.8
13.	Acrux (Alpha Crucis)	0.77c	321
14.	Aldebaran (Alpha Tauri)	0.87	65
15.	Spica (Alpha Virginis)	0.98	262
16.	Antares (Alpha Scorpii)	1.06v	604
17.	Pollux (Beta Geminorum)	1.16	34
18.	Fomalhaut (Alpha Piscis Austrini)	1.17	25.1
19.	Becrux (Beta Crucis)	1.25	352
20.	Deneb (Alpha Cygni)	1.25	1467
21.	Regulus (Alpha Leonis)	1.36	78
22.	Adhara (Epsilon Canis Majoris)	1.50	431
23.	Castor (Alpha Geminorum)	1.58c	52
24.	Gacrux (Gamma Crucis)	1.59cv	88
25.	Shaula (Lambda Scorpii)	1.62	359

c Combined magnitude of a double star.
v Variable by more than 0.2 magnitude.

APPENDIX 3—THE CONSTELLATIONS

Name	Genitive	Abbreviation
Andromeda	Andromedae	And
Antlia	Antliae	Ant
Apus	Apodis	Aps
Aquarius	Aquarii	Aqr
Aquila	Aquilae	Aql
Ara	Arae	Ara
Aries	Arietis	Ari
Auriga	Aurigae	Aur
Boötes	Boötis	Boo
Caelum	Caeli	Cae
Camelopardalis	Camelopardalis	Cam
Cancer	Cancri	Cnc
Canes Venatici	Canum Venaticorum	CVn
Canis Major	Canis Majoris	CMa
Canis Minor	Canis Minoris	CMi
Capricornus	Capricorni	Cap
Carina	Carinae	Car
Cassiopeia	Cassiopeiae	Cas
Centaurus	Centauri	Cen
Cepheus	Cephei	Cep
Cetus	Ceti	Cet
Chamaeleon	Chamaelontis	Cha
Circinis	Circini	Cir
Columba	Columbae	Col
Coma Berenices	Comae Berenices	Com
Corona Australis	Coronae Australis	CrA
Corona Borealis	Coronae Borealis	CrB
Corvus	Corvi	Crv
Crater	Crateris	Crt
Crux	Crucis	Cru
Cygnus	Cygni	Cyg
Delphinus	Delphini	Del
Dorado	Doradus	Dor
Draco	Draconis	Dra
Equuleus	Equulei	Equ

APPENDIX 3—THE CONSTELLATIONS (Continued)

Eridanus	Eridani	Eri
Fornax	Fornacis	For
Gemini	Geminorum	Gem
Grus	Gruis	Gru
Hercules	Herculis	Her
Horologium	Horologii	Hor
Hydra	Hydrae	Hya
Hydrus	Hydri	Hyi
Indus	Indi	Ind
Lacerta	Lacertae	Lac
Leo	Leonis	Leo
Leo Minor	Leonis Minoris	LMi
Lepus	Leporis	Lep
Libra	Librae	Lib
Lupus	Lupi	Lup
Lynx	Lyncis	Lyn
Lyra	Lyrae	Lyr
Mensa	Mensae	Men
Microscopium	Microscopii	Mic
Monoceros	Monocerotis	Mon
Musca	Muscae	Mus
Norma	Normae	Nor
Octans	Octantis	Oct
Ophiuchus	Ophiuchi	Oph
Orion	Orionis	Ori
Pavo	Pavonis	Pav
Pegasus	Pegasi	Peg
Perseus	Persei	Per
Phoenix	Phoenicis	Phe
Pictor	Pictoris	Pic
Pisces	Piscium	Psc
Piscis Austrinus	Piscis Austrini	PsA
Puppis	Puppis	Pup
Pyxis	Pyxidis	Pyx
Reticulum	Reticuli	Ret
Sagitta	Sagittae	Sge
Sagittarius	Sagittarii	Sgr

APPENDIX 3—THE CONSTELLATIONS (Continued)

Scorpius	Scorpii	Sco
Sculptor	Sculptoris	Scl
Scutum	Scuti	Sct
Serpens	Serpentis	Ser
Sextans	Sextantis	Sex
Taurus	Tauri	Tau
Telescopium	Telescopii	Tel
Triangulum	Trianguli	Tri
Triangulum Australe	Trianguli Australis	TrA
Tucana	Tucanae	Tuc
Ursa Major	Ursae Majoris	UMa
Ursa Minor	Ursae Minoris	UMi
Vela	Velorum	Vel
Virgo	Virginis	Vir
Volans	Volantis	Vol
Vulpecula	Vulpeculae	Vul

APPENDIX 4: THE MESSIER OBJECTS

Abbreviations: SNR (supernova remnant), GC (globular cluster), OC (open cluster), DN (diffuse nebula), PN (planetary nebula), SG (spiral galaxy), EG (elliptical galaxy), IR (irregular galaxy)

No.	Constellation	Size	Magnitude*	Type	Name
1	Tau	6 × 4	8.4 (8.0)	SNR	Crab Nebula
2	Aqr	13	6.6 (6.3)	GC	
3	CVn	16	6.3 (5.9)	GC	
4	Sco	26	5.4	GC	
5	Ser	17	5.7	GC	
6	Sco	15	4.2	OC	
7	Sco	80	3.3 (2.8)	OC	
8	Sgr	90 × 40	4.6 (3.3)	DN	Lagoon Nebula
9	Oph	9	7.8	GC	
10	Oph	15	6.6	GC	
11	Sct	14	5.8 (5.3)	OC	Wild Duck Cluster
12	Oph	14	6.1 (6.8)	GC	
13	Her	17	5.8 (5.3)	GC	
14	Oph	12	7.6	GC	
15	Peg	12	6.3 (6.0)	GC	
16	Ser	120 × 25	6.0 (cluster)	OC/DN	Eagle Nebula
17	Sgr	46 × 37	6.0 (cluster)	DN/OC	Omega Nebula
18	Sgr	9	6.9	OC	
19	Oph	14	6.8	GC	
20	Sgr	29 × 27	6.3 (cluster)	DN/OC	Trifid Nebula
21	Sgr	13	5.9	OC	
22	Sgr	24	5.2	GC	
23	Sgr	27	5.5	OC	
24	Sgr	120 × 60	4.6 (2.5)		Small Sagittarius Star Cloud
25	Sgr	32	4.6	OC	
26	Sct	15	8.0	OC	
27	Vul	8 × 6	7.3	PN	Dumbbell Nebula
28	Sgr	11	6.9	GC	
29	Cyg	7	6.6	OC	
30	Cap	11	6.9	GC	

APPENDIX 4: THE MESSIER OBJECTS (Continued)

No.	Constellation	Size	Magnitude*	Type	Name
31	And	178 × 63	3.4	SG	Andromeda Galaxy
32	And	8 × 6	8.2	EG	
33	Tri	62 × 39	5.7	SG	Pinwheel Galaxy
34	Per	35	5.2	OC	
35	Gem	28	5.1	OC	
36	Aur	12	6.0	OC	
37	Aur	24	5.6	OC	
38	Aur	21	6.4	OC	
39	Cyg	32	4.6	OC	
40	UMa		8	Double star Winnecke 4, mags. 9.0, 9.6	
41	CMa	38	4.5	OC	
42	Ori	66 × 60	3.7	DN	Great Orion Nebula
43	Ori	20 × 15	6.8	DN	
44	Cnc	95	3.1	OC	The Beehive
45	Tau	110	1.5	OC	The Pleiades
46	Pup	27	6.1	OC	
47	Pup	30	4.4 (5.7)	OC	
48	Hya	54	5.8	OC	
49	Vir	9 × 7	8.4	EG	
50	Mon	16	5.9	OC	
51	CVn	11 × 8	8.4	SG	Whirlpool Galaxy
52	Cas	13	6.9	OC	
53	Com	13	7.7	GC	
54	Sgr	9	7.7 (7.2)	GC	
55	Sgr	19	6.3	GC	
56	Lyr	7	8.4	GC	
57	Lyr	1	8.8	PN	Ring Nebula
58	Vir	5 × 4	9.6	SG	
59	Vir	5 × 3	9.6	EG	
60	Vir	7 × 6	8.8	EG	
61	Vir	6 × 5	9.6	SG	
62	Oph	14	6.4 (6.7)	GC	

APPENDIX 4: THE MESSIER OBJECTS (Continued)

No.	Constellation	Size	Magnitude*	Type	Name
63	CVn	12 × 8	8.6	SG	
64	Com	9 × 5	8.5	SG	Black-eye Galaxy
65	Leo	10 × 3	9.3 (8.8)	SG	
66	Leo	9 × 4	9.0	SG	
67	Cnc	30	6.9 (6.0)	OC	
68	Hya	12	7.3 (7.6)	GC	
69	Sgr	7	7.7 (7.4)	GC	
70	Sgr	8	7.8	GC	
71	Sge	7	8.4 (8.0)	GC	
72	Aqr	6	9.2	GC	
73	Aqr	1	8.9		Small group of four stars
74	Psc	10 × 9	9.4 (8.5)	SG	
75	Sgr	6	8.6	GC	
76	Per	2 × 1	10.1	PN	Little Dumbbell Nebula
77	Cet	7 × 6	8.9	SG	
78	Ori	8 × 6	(8.0)	DN	
79	Lep	9	7.7	GC	
80	Sco	9	7.3	GC	
81	UMa	26 × 14	6.9	SG	
82	UMa	11 × 5	8.4	IG	
83	Hya	11 × 10	7.5	SG	
84	Vir	5 × 4	9.1	EG	
85	Com	7 × 5	9.1	EG	
86	Vir	7 × 6	8.9	EG	
87	Vir	7	8.6	EG	
88	Com	7 × 4	9.6	SG	
89	Vir	4	9.7	EG	
90	Vir	10 × 5	9.5	SG	
91	Com	5 × 4	10.1	SG	
92	Her	11	6.5	GC	
93	Pup	22	6.2	OC	
94	CVn	11 × 9	8.2	SG	
95	Leo	7 × 5	9.7	SG	
96	Leo	7 × 5	9.2	SG	

APPENDIX 4: THE MESSIER OBJECTS (Continued)

No.	Constellation	Size	Magnitude*	Type	Name
97	UMa	3	9.9	PN	Owl Nebula
98	Com	10 × 3	10.1	SG	
99	Com	5	9.9	SG	
100	Com	7 × 6	9.3	SG	
101	UMa	27 × 26	7.9	SG	
102				Duplicate observation of M101	
103	Cas	6	7.4	OC	
104	Vir	9 × 4	8.0	SG	Sombrero Galaxy
105	Leo	4 × 4	9.3	EG	
106	CVn	18 × 8	8.3	SG	
107	Oph	10	7.8	GC	
108	UMa	8 × 2	10.0	SG	
109	UMa	8 × 5	9.8	SG	
110	And	17 × 10	8.0	EG	

*Dimensions are from *Sky Catalogue 2000.0*, as are magnitudes, except those in parentheses, which are from Steve O'Meara's *The Messier Objects*.

APPENDIX 5: Positions of Other Celestial Objects

Here are positions of objects mentioned in the text which for one reason or other are not plotted on this book's maps. The objects are listed in order of their appearance in the text. The figures for R.A. and declination given are epoch 2000 coordinates.

	R.A.	Declination
December		
The Double Cluster of Perseus:		
NGC 869 (h Persei)	2h19m	+57° 09′
NGC 884 (Chi Persei)	2h 22m	+57° 07′
Alpha Persei Cluster	3h 22m	+49°
NGC 1499 (California Nebula)	4h 01m	+36° 37′
32 Eridani	3h 54m	−02° 57′
January		
NGC 1977	5h 36m	−04° 52′
NGC 2024	5h 41m	−02° 27′
W Orionis	5h 05m	+01 11′
AE Aurigae 5h 16m	+34° 19′	
R Leporis (Hind's Crimson Star)	5h 00m	−14° 48′
February		
NGC 2362		
(Tau Canis Majoris Cluster)	7h 19m	−24° 57′
h3945 (145 Canis Majoris)	7h 17m	−23° 19′
NGC 2244		
(Rosette Nebula's cluster)	6h 32m	+04° 52′
NGC 2264		
(Christmas Tree Cluster)	6h 41m	+09° 53′
NGC 2261		
(Hubble's Variable Nebula)	6h 39m	+08° 44′
NGC 2392 (Eskimo Nebula)	7h 29m	+20° 55′

March

R Leonis	9h 48m	+11° 26´
NGC 2903	9h 32m	+21° 30´
Zeta Cancri	8h 12m	+17° 39´
12 Lyncis	6h 46m	+59° 27´
19 Lyncis	7h 23m	+55° 17´
38 Lyncis	9h 19h	+36° 48´
NGC 2419		
(Intergalactic Wanderer)	7h 38m	+38° 53´
NGC 2683	8h 53m	+33° 25´
NGC 3115	10h 05m	−07° 43´
NGC 3242 (Ghost of Jupiter)	10h 25m	−18° 38´
U Hydrae	10h 38m	−13° 23´
NGC 2477	7h 52m	−38° 33´
L2 Puppis	7h 14m	−44° 39´
NGC 3132 (Eight-Burst Nebula)	10h 08m	−40° 26´

April

NGC 4631		
(Humpback Whale Galaxy)	12h 42m	+32° 32´
NGC 4244	12h 18m	+37° 49´
NGC 4361	12h 24m	−18° 48´
NGC 4038-4039 (Ring-Tail		
Galaxy, Antennae)	12h 02m	−18° 52´

May

44 Bootis	15h 04m	+47° 39´
Struve 1835	14h 23m	+08° 27´
3C273	12h 29m	+02° 03´
NGC 4725	12h 50m	+25° 30´
NGC 4565	12h 36m	+25° 59´
Coma Star Cluster	12h 25m	+26°
NGC 5139 (Centaurus A)	13h 26m	−43° 01´

June

R Coronae Borealis	15h 49m	+28° 09´
T Coronae Borealis (the Blaze Star)	16h 00m	+25° 55´
Zeta Coronae Borealis	15h 39m	+36° 38´
Sigma Coronae Borealis	16h 15m	+33° 52´
Eta Coronae Borealis	15h 23m	+30° 17´
95 Herculis	18h 02m	+21° 36´
70 Ophiuchi	18h 06m	+02° 30´
IC 4665	17h 46m	+05° 43´
Barnard's Star	17h 58m	+04° 34´
17-16 Draconis	16h 36m	+52° 55´
41-40 Draconis	18h 00m	+80° 00´
NGC 6543 (Cat's Eye Nebula)	17h 59m	+66° 38´

July

Struve 2470/2474 (the Double Double's Double)	19h 09m	+34° 31´
NGC 6709	18h 52m	−10° 21´
36 Ophiuchi	17h 15m	−26° 36´
NGC 6231	16h 54m	−41° 48´

August

61 Cygni	21h 07m	+38° 45´
16 Cygni	19h 42m	+50° 32´
NGC 6826 (the Blinking Planetary)	19h 45m	+50° 31´
P Cygni	20h 18m	+38° 02´
M11	18h 51m	-06° 16´
S-O Double Cluster:		
NGC 6633	18h 28m	+06° 34´
IC 4756	18h 39m	+05° 27´

September

NGC 6940	20h 35m	+28° 18′
The Coathanger (Brocchi's		
Cluster, Collinder 399)	19h 25m	+20° 1′

October

NGC 7331	22h 37m	+34° 25′
NGC 7209	22h 05m	+46° 30′
8 Lacertae	22h 36m	+39° 38′
TX Piscium	23h 46m	+03° 29′
NGC 253	00h 48m	−39° 11′

November

NGC 7521/56 Andromedae	01h 58m	+37° 50′
Rho Cassopeiae	23h 54m	+57° 30′
NGC 457 (ET or Owl Cluster		
with Phi Cassiopeiae)	01h 19m	+58° 20′
NGC 663	01h 46m	+61° 15′
NGC 7789	23h 57m	+56° 44′
NGC 1502 (Golden Harp Cluster,		
with Kemble's Cascade)	04h 08m	+62° 20′
NGC 2403	07h 37m	+65° 36′

Notes

1. Paul Kunitzsch and Tim Smart, *Short Guide to Modern Star Names and Their Derivations* (Wiesbaden, Germany: Harrassowitz, 1986), pp. 45–46. In 2001, this book also became available from Sky Publishing Corporation.

2. Robert Burnham Jr., *Burnham's Celestial Handbook*, vol. 2 (New York: Dover Publications, 1978), p. 1202.

3. James Mullaney, *Celestial Harvest* (New York: Dover Publications, 2002), p. 10.

4. Ibid., p. 23.

5. Robert Burnham Jr., *Burnham's Celestial Handbook*, vol. 3 (New York: Dover Publications, 1978), p. 2068.

6. Guy Ottewell, *Astronomical Calendar 2002* (Middleburg, Va.: Universal Workshop, 2001), p. 33.

7. Stephen James O'Meara, *The Messier Objects* (Cambridge, Mass. and Cambridge, England: Sky Publishing Corporation and Cambridge University Press, 1998), p. 290.

8. Robert Burnham Jr., *Burnham's Celestial Handbook*, vol. 2 (New York: Dover Publications, 1978), p. 768.

9. James Mullaney, *Celestial Harvest* (New York: Dover Publications, 2002), p. 24.

10. Walter Scott Houston, quoted by Robert Burnham Jr., *Burnham's Celestial Handbook*, vol. 3 (New York: Dover Publications, 1978), p. 1752.

11. Stephen James O'Meara, *The Messier Objects* (Cambridge, Mass. and Cambridge, England: Sky Publishing Corporation and Cambridge University Press), p. 87.

12. J. R. R. Tolkien, *The Hobbit* (Boston: Houghton Mifflin Company, 1966), p. 283.

13. Mullaney, *Celestial Harvest*, p. 26.

14. Ibid., p. 92.

15. R. H. Allen, *Star Names—Their Lore and Meaning* (New York: Dover Publications, 1963), p. 39.

16. Willy Ley, *Watchers of the Skies* (New York: The Viking Press, 1969), p. 183.

17. George Robert Kepple and Glen W. Sanner, *The Night Sky Observer's Guide*, vol. 1 (Richmond, Va.: Willmann-Bell, Inc., 1999).

Glossary

absolute magnitude. The magnitude of brightness a star would have if seen at a standard distance of 10 parsecs (about 32.6 light-years).

altazimuth system. System for indicating positions in the sky using altitude and azimuth as vertical and horizontal measures.

altitude. Apparent angular height in the sky (vertical measurement in the altazimuth system).

apparent magnitude. The magnitude of brightness a star appears to have in our sky.

asterism. A pattern of stars in the sky which is not an official constellation.

atmospheric extinction. Dimming of the light of celestial objects due to absorption and scattering by Earth's atmosphere.

aurora. Patterns of radiance (usually moving and fluctuating) produced when atomic particles from the Sun are energized in Earth's magnetic field and channeled to collide with upper-atmosphere gases in regions surrounding Earth's magnetic poles. Also known as the Northern Lights or aurora borealis (aurora australis in southern hemisphere).

averted vision. Technique of looking slightly to the side of a faint object in order to increase its visibility by allowing its light to fall on the parts of the eye's retina most sensitive to light.

azimuth. Horizontal measure around the sky in the altazimuth system.

binary star. A double-star system in which the members are believed to be orbiting around each other (that is, around a common center of gravity).

black hole. An object, thought to be the result of a massive star's collapse, whose gravity has become so strong as to prevent even light (and other electromagnetic radiations) from escaping from it.

blue giant. A massive, large, and extremely luminous star with a very high surface temperature, which causes it to appear bluish white to the eye.

celestial sphere. The imaginary sphere surrounding Earth with an inner surface that is all the sky both above and below one's horizon.

Cepheid. A type of pulsating variable star whose precisely regular brightness variations can be employed to estimate the star's distance by use of the period-luminosity relation (if you know the period of the Cepheid's brightness variations, you know its luminosity—true brightness—and can therefore compare this to its apparent brightness to determine distance).

circumpolar. Close enough to one of the celestial poles so as to never rise or set but rather circle around the pole above the horizon.

comet. A mostly icy body, far less massive than the planets, that produces a cloud of dust and/or gas (coma) when it is heated sufficiently by the Sun (or, less commonly, by other forces).

conjunction. Strictly speaking, the arrangement when one celestial object moves to have the same right ascension or ecliptic longitude (passes due north or due south of a second object in a celestial coordinate system) of another. More loosely, any temporary pairing or gathering of celestial objects which is considered close.

constellation. An official pattern of stars or, more strictly, the officially demarcated section of sky in which that pattern lies.

crescent. Phase in which a world's hemisphere that is facing us appears less than half-lit.

culminate. Reach the north-south meridian of the sky, typically achieving highest point above horizon.

dark adaptation. The increase in the sensitivity of our eyes to dim light that occurs when they are kept away from bright light for a while.

dark nebula. A nebula that does not shine by either emitted or reflected light and is therefore visible only in silhouette against a more distant bright nebula or starry background.

declination. North-south measure in the equatorial system of celestial coordinates, corresponding to latitude on Earth.

deep-sky object. An object beyond our solar system, though the term is usually not applied to individual stars or double- and multiple-star systems but rather to star clusters, nebulas, and galaxies.

diffuse nebula. A luminous nebula that shines from reflecting the light of nearby stars (reflection nebula) or is heated enough by very hot stars to glow on its own (emission nebula).

diurnal motion. The apparent motion of celestial objects caused by the Earth's rotation.

double star. A star that, upon closer or more sophisticated examination, turns out to consist of two or more component stars.

eclipse. The hiding or dimming of one object by another object or by another object's shadow.

eclipsing binary. A type of variable star in which one component star of a double star eclipses the other, or both alternately eclipse each other, causing the variations in brightness.

ecliptic. The apparent path of the Sun through the zodiac constellations, which is really the projection of Earth's orbit in the sky.

equatorial system. A system for indicating positions in the heavens using right ascension (corresponding to longitude on Earth) and declination (corresponding to latitude on Earth).

fireball. A meteor brighter than Venus.

full-cutoff fixture. A light fixture which emits light entirely below the horizontal, eliminating directly produced skyglow and reducing light pollution.

galaxy. An immense congregation of typically billions of stars forming a system of spiral, elliptical, or irregular shape.

galactic cluster. *See* "open cluster."

gibbous. Phase in which the hemisphere of a world facing us is more than half-lit but less than fully lit.

globular cluster. A kind of star cluster consisting of tens of thousands up to a few million stars arranged in a roughly spherical shape.

horizon. The boundary line between sky and land or sea.

H-R diagram (Hertzsprung-Russell diagram). A tremendously revealing diagram that plots the true brightness (in terms of absolute magnitude or luminosity) of stars versus their spectral class or surface temperature.

light pollution. Excessive or misdirected artificial outdoor lighting.

light-year. The distance which light, fastest thing in the universe, travels in the course of one year.

limb. The edge of the Moon or other celestial body.

long-period variable. A major kind of variable star, in which the period of brightness variations is months or years long and the range of the variations typically great, with the amount of both often being irregular or only semiregular.

luminosity. The true brightness of a star, independent of its distance from us, measured in units of the Sun's true brightness (for instance, a star twice as luminous as the Sun has a luminosity of 2).

magnitude. A measure of brightness in astronomy, in which an object one hundred times brighter than another is exactly five magnitudes brighter. The brighter the object, the lower the magnitude figure (e.g., a 1st-magnitude star is brighter than a 2nd-magnitude star), with negative magnitudes for the very brightest objects of all.

meridian. The line in the sky that extends from the due south horizon to the zenith onward to the due north horizon.

Messier objects. Also known as M-objects, one of slightly more than one hundred deep-sky objects cataloged in the eighteenth century by French astronomer Charles Messier.

meteor. The streak of light seen when a piece of space rock or iron enters Earth's atmosphere and burns up (due to, most roughly put, friction).

meteorite. A piece of space rock or iron which survives its trip through the atmosphere as a meteor to reach the ground and be found.

meteoroid. A rocky or metallic natural object smaller than an asteroid (no more than a few hundred meters across, probably very much smaller, even dust-size) which would become a meteor if it entered Earth's atmosphere and a meteorite if it reached Earth's surface.

meteor shower. An increased number of meteors all appearing to diverge from the direction of a single area among the constellations.

meteor storm. An extremely intense meteor shower, in which rates of more than one thousand meteors per hour may be observed.

meteor train. The trail of ionization left by some meteors to linger glowing after the meteor itself has disappeared.

Milky Way. The great spiral galaxy in which we ourselves live, and also the night sky's band of strongest glow from the combined light of innumerable distant stars in the galaxy's equatorial plane.

moon. A rocky or icy object which circles a planet. Also known as a (natural) satellite.

multiple star. A star system consisting of more than two stars (although "double star" is often used as the umbrella term for systems of two, three, four, etc., stars anyway).

nebula. A vast cloud of dust and gas in interstellar space. Different types include diffuse nebulas (which include emisssion nebulas and reflection nebulas), planetary nebulas, and dark nebulas.

neutron star. The collapsed, ultradense core of a star left after a supernova formed from an original star not massive enough to collapse all the way into becoming a black hole.

Northern Lights. *See* "aurora."

nova. An exploding (and therefore briefly very much brightened) star that loses a small fraction of its mass in the outburst, which may often arise from interactions between stars in double-star systems.

occultation. The hiding of one celestial object by another (sometimes such an event is instead, or at least primarily, called an eclipse).

open cluster. Also called a galactic cluster, this major kind of star cluster consists of a usually irregular shape and includes typically dozens or a few hundred stars.

optical double. A double star in which the two components are not truly related, one object being much farther away and just happening to lie on nearly the same line of sight as seen from Earth.

P.A. See "position angle."

parallax. The change in a star's position caused by our change in viewpoint (usually our change from one side of Earth's orbit to the other.

parsec. A parallax second, the distance at which the view from opposite sides of Earth's orbit would cause an object to have an apparent position change (a parallax) of 1 arc-second (1 parsec is equal to about 3.26 years).

planet. A relatively massive world (but not massive enough to sustain thermonuclear reactions like a star) in direct orbit about a star.

planetary nebula. A cloud of gas and dust cast off by a hot, small, dying white dwarf star. (The name comes from the passing resemblance of some of these blue or green nebulas to planets like Uranus and Neptune as seen in the telescope.)

position angle (P.A.). The direction angle of a companion star in relation to its primary in a double-star system.

precession. A slight wobble in the rotational axis of the Earth, caused by the pulls of the other solar system bodies, and resulting in slow changes of the direction of the north celestial pole and other positions in the heavens.

proper motion. The motion of a star relative to the sun as projected on the celestial sphere—in other words, the change in a star's position on the celestial sphere produced by the component of its space velocity (motion through space) which is transverse (neither toward nor away from us).

pulsar. A type of neutron star oriented toward Earth in such a way that we get to observe pulses of light and sometimes of other eletromagnetic wavelengths released from gaps in the exploding star's magnetic field near its poles.

quasar. An incredibly powerful source of light and other electromagnetic wavelengths which may be a kind of intense core of a galaxy.

R.A. *See* "right ascension."

radiant. The region in the sky from which the meteors of a shower all appear to diverge.

red giant. A huge and fairly massive star of extremely low density which radiates mostly in the red due to its relatively low surface temperature.

red dwarf. A small star far less massive than the Sun which radiates relatively little light and mostly in the red due to its comparatively low surface temperature.

revolution. The orbiting of one celestial body around another (e.g., the Earth's revolution period around the Sun is one year).

right ascension. (R.A.). West-east measure in the equatorial system of celestial coordinates, corresponding to longitude on Earth (though expressed somewhat differently than longitude on Earth: in hours of right ascension from 0 to 24).

rotation. The spinning of a celestial object (e.g., the Earth's rotation period is one day).

satellite. Any celestial body which orbits another, but in practice usually confined to a body which orbits a planet. An artificial satellite is one launched by humankind to orbit Earth or another world; a natural satellite (composed of rock and/or ice) is more popularly known as a moon.

"seeing." Sharpness of astronomical images as a function of turbulence in Earth's atmosphere.

sidereal time. Time measured by the passage of the stars around the sky, without reference to the Sun (the sidereal day is about four minutes shorter than the solar day).

skyglow. The component of light pollution (excessive and misdirected artificial outdoor lighting) which goes up into the sky. Around even just fairly large cities, it is visible for dozens of miles.

SNR. *See* "supernova remnant."

spectral class. The subcategory of spectral type in which a star belongs—for instance, O3, G2, M5, in which the letter denotes the type and the number indicates the class.

spectral type. The category—ranging from O (the hottest normal stars) to M (the coolest normal stars), with some alternative unusual types—into which a star is placed according to the appearance of its light's chemical spectrum.

solar system. The whole collection of planets, moons, asteroids, comets, and meteoroids orbiting the Sun, or another star, under the Sun's or other star's gravitational influence.

sporadic meteor. A meteor not traceable to any known meteor shower.

star. A massive self-luminous ball of gas producing energy by nuclear fusion in a dense core (also called a sun or, in the case of Earth's star, the Sun).

star cluster. A grouping of anywhere from a few to several million stars traveling through space relatively close together but not closely enough to be considered a multiple star.

supernova. The much more powerful kind of star collapse, explosion and brightening, in which a large part of a star's mass is lost, and the star's core may become a neutron star or black hole.

supernova remnant (SNR). The cloud of material ejected by a supernova, sometimes visible for many thousands of years afterward.

transparency. The degree to which Earth's atmosphere is capable of letting celestial light pass through it. (How clear of dust and water vapor is the atmosphere over you tonight?)

Universal Time. (UT). Time system for dating astronomical events, corresponding to the local time at Greenwich, England, on the 0° meridian of longitude on Earth (to obtain Universal Time from your current local standard time, add five hours to Eastern Standard Time, four hours to Central Standard Time, and so on).

UT. *See* "Universal Time."

variable star. A star which, for one of a number of possible reasons, undergoes repeated changes in its brightness.

white dwarf. A hot, extremely small but fairly massive and therefore very dense star that represents the last luminous stage in the life of many stars.

zenith. The overhead point in the sky.

zodiac. The circle of constellations through which the Sun passes during the course of the year.

Sources of Information

BOOKS

Burnham, Jr., Robert. *Burhham's Celestial Handbook*. 3 vols. New York: Dover Publications. The classic and still best all-around guide to most deep-sky sights visible through telescopes up to about 10-inch aperture. Some of the information from this 1978 work is outdated, but it remains a grand collection of engagingly (sometimes even poetically) written text, tables of information, and photographs about the lore, science, and especially the observational appearances of thousands of stars and deep-sky objects.

Dickinson, Terence and Alan Dyer. *The Backyard Astronomer's Guide*. Camden East, Ont.: Camden House Publishing. Very insightful and comprehensive coverage of what amateur astronomers need to know about telescopes, equipment, techniques, and observing sites.

Harrington, Phil. *Touring the Universe Through Binoculars*. New York: John Wiley & Sons. An excellent guide to selection and use of binoculars. The majority of the book is devoted to a constellation-by-constellation tour of binocular sights.

Houston, Walter Scott. *Deep-sky Wonders*. Cambridge, Mass.: Sky Publishing Corporation. Houston was the father of deep-sky observing, influencing all the other great observers of the second half of the twentieth century. His widest influence came through nearly half a century of his monthly "Deep-sky Wonders" columns in *Sky & Telescope*. This book is a rich selection of those columns, chosen and edited together by the knowledgeable Steve O'Meara. Houston's information, perspectives, practical advice, beautiful writing, and inspiration remain invaluable.

Kepple, George Robert and Glen Sanner. *The Night Sky Observer's Guide*. 2 vols. Richmond, Va.: Willmann-Bell, Inc. These superb volumes concentrate on providing descriptions, maps, photos, sketches, and recent information on more than five thousand deep-sky objects, including descriptions of what can be seen of each specifically with different sizes of telescopes all the way up to the largest (twenty-inch-plus) amateur telescopes. Using the observations of not just the authors but numerous other experienced observers, this book in many ways extends and updates Burnham's *Celestial Handbook* for advanced amateur astronomers at the turn of the twenty-first century. The authors are humble and acknowledge (rightly) that the book is not a replacement for Burnham's, however, for the latter contains a vast amount of material (about southern hemisphere objects, lore, and much more) which is outside the scope of the former.

Mullaney, James. *Celestial Harvest*. New York: Dover Publications. One of the world's most veteran deep-sky observers offers splendid and information-packed descriptions and quotes, along with data about more than three hundred of the finest deep-sky objects, including many undeservedly little-known double stars.

O'Meara, Stephen James. *The Messier Objects*. Cambridge, Mass. and Cambridge, England: Sky Publishing Corporation. One of the world's most visually gifted observers presents the 110 Messier objects in detailed and often poetic descriptions along with a photgraph, sketch, and more for each object.

Ottewell, Guy. *The Astronomical Companion*. Ottewell's seventy-two-page atlas-size guide to astronomical topics which are not year-dependent (see his Astronomical Calendar below for those that are), including dozens of huge and unique diagrams. Go to www.universalworkshop.com to order).

Schaaf, Fred. *40 Nights to Knowing the Sky*. New York: Henry Holt and Company. Comprehensive introduction to the workings and appearances of the Moon, Sun, planets, stars, and deep-sky objects on the celestial sphere and in the universe.

Schaaf, Fred. *Seeing the Sky, Seeing the Solar System,* and *Seeing the Deep Sky.* New York: John Wiley & Sons. A trilogy of books about observing projects, focusing on naked-eye observations, telescopic observations of solar system objects, and telescopic observations of objects beyond the solar system.

Schaaf, Fred. *Wonders of the Sky.* New York: Dover Publications. Introductory guide to naked-eye skywatching, including daytime sights of atmospheric optics like rainbows and halos.

ATLASES AND PLANISPHERES

Star atlases range from those displaying only stars down to fifth or sixth magnitude (a few thousand stars) to *The Millenium Star Atlas* (well over 1 million stars, all stars brighter than magnitude 11.0). A fine selection of these atlases is described in the Sky Publishing Corporation catalog (see address, web address, and phone number for *Sky & Telescope* magazine, below).

Various sky simulation software programs offer some access to maps of even dimmer stars with your computer. Descriptions of many of these can also be found in the Sky Publishing catalog, and elsewhere (catalogs of Astronomical Society of the Pacific and Orion Telescope and Binocular Center, for instance).

If you need star-finding information that you can carry around and out to your observing site easily, try a planisphere, a handheld rotating carboard or plastic rotating display of the starry sky that can be set for any time of any night. The best all-around planisphere is probably the "Precision Planet and Star Locator," produced by David Kennedal and distributed in the North America by Sky Publishing Corporation.

PERIODICALS, ALMANACS, ANNUAL GUIDES

Astronomical Calendar. Eighty-two atlas-size pages filled with original diagrams and rich text about the year's celestial events (includes several sections and month-by-month "Observers' Highlights" by Fred Schaaf, produced by Guy Ottewell at Universal Workshop (see Web address above).

Astronomy Magazine. Kalmbach Publishing, P.O. Box 1612, Waukesha, Wisc. 53187. www.astronomy.com. One of the two largest and most popular astronomy magazines, with much information about buying and learning to use telescopes.

Mercury. Fine magazine published by Astronomical Society of the Pacific. (See address in section on Audiovisual Materials and Software, below.)

Sky & Telescope. P.O. Box 9111, Belmont, Mass. 02178. www.SkyandTelescope.com. Includes planets and stars columns (plus other pieces) by Fred Schaaf plus informative articles on telescope selection and use.

Sky Calendar. Abrams Planetarium, Michigan State University, East Lansing, Mich. 48824. (See also Skywatcher's Diary at www.pa. msu.edu/abrams/diary.html.) For just $10 a year you get a two-sided sheet for each month with a basic star map on one side and informative calendar with sky scene for each night on the other side.

SkyNews. Canadian magazine shorter than *Sky & Telescope* and *Astronomy* and published bimonthly but very beautifully illustrated and offering some excellent observational articles. www.skynewsmagazine.com.

ONLINE SOURCES

American Association of Variable Star Observers. www.aavso.org.

Astronomical League. Confederation of more than 200 amateur astronomy clubs in U.S. www.mcs.net/~bstevens/al.

AstroAlerts. Available from the *Sky & Telescope* Web site (see below), this is a free e-mail service that notifies subscribers about urgently time-sensitive astronomical events that are in progress. A person can subscribe to all the AstroAlerts or just to one that pertains to solar/auroral activity or comets or variable stars or near-Earth asteroid passages, etc.

Astronomical Society of the Pacific. www.astrosociety.org.

Astronomy Magazine Online. Home page of *Astronomy* magazine. www.astronomy.com.

International Dark-sky Association. Central bureau for information on light pollution and how to combat it, technologically and legally. www.darksky.org.

International Meteor Organization. www.imo.net.

Sky & Telescope Online. Home page of *Sky & Telescope* magazine and Sky Publishing Corporation. www.SkyandTelescope.com.

Space Weather Bureau. Latest predictions, information, and images about auroras and solar activity—and about a wonderful assortment of other astronomical objects and events. www.spaceweather.com.

AUDIOVISUAL MATERIALS AND SOFTWARE

Astronomical Society of the Pacific. 390 Ashton Ave., San Francisco, Calif. 94112. This century-old educational organization offers a catalog with a wide variety of excellent books, posters, slide sets, computer programs, and CD-ROMs.

Sky Publishing Corporation. P.O. Box 9111, Belmont, Mass. 02178. The publishers of *Sky & Telescope* offer a catalog with books, star atlases, computer programs, CD-ROMs, slide sets, posters, and even globes (including ones of other planets).

Stars, Deep-sky Objects, and Constellations Index

Constellations are given in all capital letters. If a star is known well enough by its proper name, it is not also listed under its constellation by its Greek letter or other designation. Mythological characters are not indexed here, though the constellations represented by them are.

General Index

Individual objects beyond the solar system are listed in their own index. Constellations which are merely mentioned in the text but not explored appear in this index, as does the Milky Way because some mentions of it are as a concept as opposed to a sight.